13-

D0370097

THE GREAT U-TURN

Edward Goldsmith was born in Paris in 1928 and was educated at Millfield and Magdalen College, Oxford. Though once involved in various commercial pursuits, he has devoted most of his adult life to study—his main interests being philosophy, anthropology, theoretical biology and ecology. In 1969 he founded *The Ecologist* of which he is still publisher and joint editor. He has written many articles which have appeared in journals and magazines around the world, and has lectured widely and taught several university courses, both in the UK and the US. He has also written, co-authored or co-edited the following books: *Can Britain Survive?* (1971), *A Blueprint for Survival* (1972), *The Stable Society* (1978), *La Medecine à la Question* (1982), *The Social and Environmental Effects of Large Dams* (2 vols, 1984, 1986), *Green Britain or Industrial Wasteland?* (1986) and *Planet Earth, Annual Report* (1988).

EDWARD GOLDSMITH

THE GREAT U-TURN

De-industrializing Society

cartoons by Richard Willson

GREEN BOOKS

© Edward Goldsmith 1988

First published by
Green Books
Hartland
Bideford
Devon, EX39 6EE

Cover: Simon Willby

British Library Cataloguing in Publication Data
Goldsmith, Edward, 1928-
The Great U-Turn:
De-industrializing Society
1. Economic development
Ecological aspects
I. Title
330.9
ISBN 1-870098-01-3

Typeset by Penwell Ltd
Parkwood, Callington, Cornwall

Printed by Hartnolls Ltd
Victoria Square, Bodmin, Cornwall

The Great U-Turn is printed on 100% recycled paper

Contents

It is difficult for me to thank all those who have helped me develop that particular world-view that is reflected in each of the chapters of this book. However, particular thanks are due to my close colleagues Peter Bunyard and Nicholas Hildyard. Thanks are also due to Richard Willson for the accompanying cartoons, some of which have already appeared over the last seventeen years in *The Ecologist*. I would like to thank Brian Price too for vetting Chapter 5, Can Pollution be Controlled? I must also thank Ruth Lumley-Smith for editing and correcting the final text and my secretary, Hilary Datchens, for organizing it (and me) and typing the various drafts.

The Fall of the Roman Empire: a social and ecological interpretation, was published in *The Ecologist*, July 1975 and in *Le Sauvage* (France) April 1976.

Education: What For? was published in *The Ecologist*, January 1974, in *PHP* (Japan) December 1975, and in *Oko Journal* (Switzerland) February 1975.

The Ecology of Unemployment was published in *The Ecologist*, February 1974 and in *Everyman's* (India) 9 February 1975.

The Ecology of Health was published in *The Ecologist*, July 1980 and as a chapter in Jean-Pierre Brunetti and Edward Goldsmith (eds) *La Medecine à la Question* (France) Editions Nathan, 1981.

Can Pollution be Controlled? was published in two instalments in *The Ecologist* Oct/Nov 1979 and Nov/Dec 1979.

The Ecology of War was published in *The Ecologist*, May 1974 and in *Le Sauvage* (France) April 1975.

De-industrializing Society was published in *The Ecologist*, May 1977, as a chapter in *Die Tagliche Revolution* (Germany) Fischer Alternativ 1978, and as an appendix to Gianfranco de Santis 'La Sociologia Ecologica di Edward Goldsmith', doctoral thesis (Tesi di Laurea) Facolta' di Magistero, University of Rome, 1977.

Foreword

EDWARD GOLDSMITH is a radical in the true sense of the word. For him, no analysis is rigorous enough unless it is pursued to its logical conclusions: no solution acceptable in the long-term unless it tackles the roots (Latin: *radices*) of the problem.

Were he dealing with some arcane philosophical riddle, or some minor point of political theory, such intellectual rigour would cause little offense: on the contrary, it would be widely applauded. But applied to the problems of modern industrial society—from pollution to degenerative disease to the miseries of social alienation—Goldsmith's 'root-and-branch' radicalism poses such a fundamental challenge to conventional thinking that many have reacted by rejecting its conclusions out of hand.

Nor should that surprise us: few of us willingly undertake to subject our most basic values to a fundamental reappraisal, for to do so inevitably threatens to undermine the very premises that underpin our whole way of life. Every 'No' to received wisdom is a 'Yes' to being an outsider—an unenviable status at the best of times.

Yet, just such a reappraisal of our basic values is inescapable if the ecological holocaust being unleashed throughout the world in the name of 'progress' is to be halted. Indeed, it is only when we examine our basic values that we begin to grasp how far we have strayed from a way of life that is compatible with the long-term health of the natural world on which we depend for our survival. For, as Goldsmith makes clear throughout this seminal book, the social and ecological problems confronting us today cannot be solved by a little tinkering with the system: they are too deep-seated, having their roots in the very nature of industrialism. Indeed (and this is a critical strand in his argument), Goldsmith rightly points out that conventional economic and technological 'solutions' often serve only to exacerbate the very problems which they are intended to solve. His discussions of the health crisis and the problems of unemployment make the point in a masterly way.

Given that Goldsmith indentifies industrialization as the root cause of today's ecological and social problems, it is only logical that he sees their solution as lying in 'de-industrialization'. Here he is at his most controversial. For unlike many political thinkers, his alternative society is modelled not on some imaginary utopia but on the real world of those traditional societies which for millennia have lived a prosperous and healthy life without destroying their environment or causing social alienation. To his critics, that emphasis on such 'tribal' societies reflects at best romantism, at worst an attempt to turn the clock back. But as Goldsmith himself points out, traditional societies are the only groups in existence which can provide us with a working model of social and ecological stability. It would be arrogant in the extreme not to learn the lessons that such societies have to offer. As for turning the clock back, that is clearly impossible. What is possible, however, and eminently desirable, is a reversal of those policies which are leading us relentlessly to destruction. The question is: will we make the great u-turn in time?

Nicholas Hildyard
March 1988

Introduction

THIS BOOK CONTAINS a series of essays written during the last sixteen years. They have been published separately (in *The Ecologist* and other periodicals and books outside the UK) but are complementary and were designed for inclusion in a single book. They have been revised and updated when this has been necessary, but no substantial changes have been made.

The first essay attempts to show that the problems we face today are not new. This is well illustrated by the fate of the Roman Empire, whose decline was not the result of the barbarian invasions but, as that of the western world today, the consequence of social disintegration and the associated environmental degradation.

If, moreover, these trends proved fatal, it is that those who governed the Empire chose—as our leaders have also chosen today—to apply short-term expedients to mask the symptoms of a disease whose fundamental cause it was not politically expedient nor economically viable to address.

The next six essays deal with a few of the symptoms of the same disease as it affects our industrial society—poor education (chapter 2), unemployment (chapter 3), ill-health (chapter 4), pollution (chapter 5) and war (chapter 6).

Others could equally well have been chosen—deforestation for instance, man-made climate change, the population explosion, malnutrition and famine, or drug-addiction and delinquency.

As I try to show in this book, our society is incapable, for a number of reasons, of solving any such problems. The first is that, in the light of the industrial world-view with which we have all been imbued, it is impossible to understand their real nature. This is partly because of the way our knowledge is compartmentalized into separate watertight disciplines, partly too because of the reductionist methodology of modern science, both of which conspire to make us see our problems in isolation from each other rather than as part of the same general picture.

It is also because of the tendency of those working within such disciplines to view the problems we face in the light of the very short experience of our industrial society, which is but a fraction of the total human experience on this planet, during much of which time such problems either did not exist or did so in a very attenuated form.

It is mainly, however, that we do not really want to understand them, for, if we did, we would have to face the unpleasant fact that they are the inevitable concomitants of those trends we most highly prize—those we have been taught to identify with 'progress'—the development of science, technology, industry, the global market system and the modern state. We would also have to face the still less acceptable conclusion that our problems can only be solved by reversing these developments, i.e. by putting 'progress' into reverse—an enterprise that few would even be willing to contemplate, yet for which there is no real alternative.

How it could be done, my colleagues and I explained in *A Blueprint for Survival* in 1972. A more detailed plan was published in the May 1977 issue of *The Ecologist* under the title of 'De-industrializing Society'. It is reproduced in the final chapter of this book. The fact that such a plan is neither politically expedient nor economically viable is no argument against it. Indeed it is increasingly apparent that if any strategy is politically expedient and economically viable, it is, as John Davoll has noted, unlikely to work.

It must follow that it is our politico-economic system itself that must be transformed, so that it becomes possible to apply real solutions to our problems rather than to persist, as we are doing today, in applying short-term expedients to mask the more apparent symptoms of these problems—a policy that, by rendering them superficially more tolerable, can only ensure their perpetuation.

The key question is what are likely to be the main features of the new society that such a transformation would bring about?

As I intimate in every chapter of this book, one can only establish this by determining what were the main features of the traditional societies of the past that proved capable, for thousands if not tens of thousands of years, of avoiding creating the terrible problems that we face today.

I regard this thesis as fundamental, for to postulate an ideal society for which there is no precedent within the human experience, as many of our political theorists, including Karl Marx, have done, is very much like postulating an alternative biology without reference to the sort of biological structures that have so far proved viable.

Clearly we cannot recreate the past, the experience of the modern age cannot be eradicated. However, in order to determine what are the necessary features of the stable, fulfilling and problem-solving societies, that, one can only hope, will emerge during the post-modern age, it is to the traditional societies of the past that we must turn for inspiration.

CHAPTER ONE

The Fall of the Roman Empire

'But in the UK, it's the barbarians who *build* our cities!'

The Fall of the Roman Empire

A SOCIAL AND ECOLOGICAL INTERPRETATION

> There is the moral of all human tales;
> 'Tis but the same rehearsal of the past;
> First Freedom, and the Glory—when that fails
> Wealth, Vice, Corruption—Barbarism at last
> And History, with all her volumes vast,
> Hath but one page.
>
> Byron

IT IS SAID of the Bourbons, that during the time they were in power, they neither learnt nor forgot anything. This could equally well be said of our political leaders, probably too, of the scientists and economists who advise them. It is a great tragedy that we seem incapable of learning the lessons of history.

Of these, one of the most instructive would undoubtedly be that of the fall of the Roman Empire. The parallel it affords with the breakdown of industrial society, which we are witnessing today, is indeed very striking. The two processes differ from one another in two principal ways. Firstly, the former was a very slow one, spread out over hundreds of years, whereas the latter is occurring at a truly frightening pace; and secondly, the role played by slavery in the former case is fulfilled in the latter by machines.

In both cases, the collapse was unexpected. In the same way that even today many intelligent people cannot bring themselves to believe that the industrial world is about to disappear for ever, the intelligentsia of Imperial Rome undoubtedly found it impossible to accept that Rome could be anything but 'eternal' and that its great civilizing influence could ever wane.

Surprising as it may seem, Rome's barbarian conquerors also seemed to share this belief. The Vandals belied their reputation and never really destroyed Rome. They had far too much respect for what it stood for. Even after Odoacer had defeated the last Western Emperor, Romulus Augustulus, and had assumed the government of the Western Empire, he carefully refrained from proclaiming himself Emperor. His letter to Emperor Zeno of Byzantium, after his victory, illustrates his great respect for the institution of the Empire. First of all as we learn from Gibbon, he tried to justify rather apologetically his abolition of the Western Empire, on the grounds that, 'The majesty of a sole monarch was sufficient to protect, at the same time, both East and West'.[1] Then he goes on to ask the Emperor to invest him with the title of 'Patrician', and with the Diocese of Italy. His successor, Theodoric, it seems, shared Odoacer's respect for the institutions of Rome.

THE BARBARIAN INVASIONS

It is customary to regard the barbarian invasions as the main cause of the fall of Rome, just as a thermonuclear war might one day mark the end of our industrial society. However, though the invasions undoubtedly contributed to the plight of Roman society, they were probably but a minor cause of its collapse.

Let us not forget that the Roman armies had been successfully fighting migrant German tribes since the days of Augustus. Why should they suddenly be overcome by those they came up against in the sixth century?

In fact, Samuel Dill considers the invasions of the third and fourth centuries to have been considerably more formidable, but, 'The invaders, however numerous,' he writes, 'are invariably driven back and in a short time there are few traces of their ravages. The truth seems to be that, however terrible the plundering bands might be to the unarmed population, yet in regular battle, the Germans were immensely inferior to the Roman troops.'[2] Ammianus, who had borne a part in many of these engagements, also points out that, in spite of the courage of the Germans, their impetuous fury was no match for the steady discipline and coolness of troops under Roman officers. The result of this moral superiority, founded on long tradition, was that the Roman soldier in the third and fourth centuries was ready to face any odds.

It would thus appear that if the invasions of the fifth and sixth centuries were more successful than the previous ones, it was not because of the increased strength of the invaders.

THE BARBARIAN RULE

If the fall of the Empire cannot be attributed to the invasions, still less can it be ascribed to the subsequent barbarian rule. From all accounts, life during this period suffered no radical change. If anything, it changed for the better under the rule of the very able Theodoric, who, according to Gibbon, re-established an age of peace and prosperity, and who did everything he could to restore the facade if not the spirit of the old Roman State, and under whom the people 'enjoyed without fear or danger the three blessings of a capital:- order, plenty and public amusements.'[3]

It must also be remembered that barbarian generals had taken over long before Odoacer defeated Romulus Augustulus, but, for reasons already mentioned, were content to remain in the background, allowing the Empire to survive under the titular head of an emperor who had some claim to legitimacy.

In fact, since the death of Theodorius, the emperors of Rome ruled in name only. During the reign of Honorius and until his murder in 408, Stilicho, a Vandal, was the effective master of the Empire. Among other things he prevented the Empire from falling to Alaric and the Goths. The Emperor

Avitus was named and supported by Theodoric. His successor, Majorian, was a nominee of the Suevian general Ricimer, as was his successor, Severus. So was the next Emperor, Anthemus, who reigned from 462 to 467. During the six years between the death of Majorian and the elevation of Anthemus, the government was entirely in the hands of Ricimer, who ruled Italy with the same independence and despotic authority which were afterwards exercised by Odoacer and Theodoric. The three emperors who followed were also the nominees of foreign powers, two of them owing their investiture to the Emperor of Byzantium, the third to a Burgundian prince. Finally, Orestes, a Pannonian, who had served in the army of Attila the Hun, took over power. He characteristically refused the purple though he accepted it in the name of his son, Romulus Augustulus, the last Emperor of the West, who was, in origin at least, a barbarian himself.

Nevertheless to refer to these men as barbarians is very misleading. As Dill points out, 'Many of these German officers were men of brilliant talents, fascinating address and noble bearing. To military skill they often added the charm of Roman culture and social patter which gave them admission even to the inner circle of the Roman Aristocracy.' Valuable testimony to this is provided by letters of the Christian, Salvianius, who passionately decried the individualism and selfishness of the ruling classes of the Empire and 'considered the Barbarians to be without question their moral superiors'.

THE FORCES WHICH MADE ROME

The fall of the Roman Empire can only be understood if we examine what were the features of Roman society which assured the success of the Republic. One can then see what were the factors that caused the erosion of these qualities during the latter days of the Republic and of the Empire.

The first of these qualities was the tribal structure of Roman society. The Roman Republic was originally made up of three separate tribes, the Ramnes, the Tities and the Luceres, each of which was in turn divided into ten clans or 'curies', of which many of the names have come down to us.[4] Indeed, the institutions of the Republic as well as the structure of its army faithfully reflected these tribal and curial divisions until the reforms of King Servius, who, like Cleisthenes in Athens, established geographical divisions to replace tribal ones as the basic administrative units of the State. Nevertheless, the tribal character of the Roman City persisted throughout the Republic.

Tribal societies are remarkably stable. Like all stable systems, they are self-regulating or self-governing. Liberty, in fact, among the Greeks meant self-government, not permissiveness as it means with us today. The Greeks were free because they ran themselves, while the Persians were slaves because they were governed by an autocrat. Self-government is only possible among a people displaying great discipline and whose cultural pattern ensures the subordination of the aberrant interests of the individual to those of the family and the society as a whole. Under such conditions, there is little need for institutions and, as has frequently been pointed out, the only institution that one finds among tribal peoples is that of the Council of Elders whose role it is to interpret tribal tradition and ensure that it is carefully observed and handed down as unchanged as possible to succeeding generations. The qualities which the classical Roman writers extolled were in fact those which are usually extolled in tribal societies.

Ennius, for instance, attributed Rome's greatness to three factors: divine favour, which presided over Rome's destiny from the very start, the steadfastness and discipline of the Romans and, finally, their moral character. This he expresses in his famous line: 'moribus antiquis res stat Romana virisque' or 'the Roman state stands firm on its ancient customs and on its men (or heroes)'.[5] It is significant that he makes no mention of the form of the Roman State as such. One can only assume that he realized that political forms are of little importance when compared with the spirit which animates them.

If, in early times, the Plebeians did not participate in government, it was because they were not members of the original tribes, nor further organized into curies, gens and families. This meant that they could not practise the religion of the Roman State which consisted of the cult of the family gods, the Lares and the Penates, of which the paterfamilias was the priest, the gods of the gens, those of the curie, the tribe and those of the City itself, of which the high priest or Pontifex Maximus was originally the king himself. An essential feature of these associated cults was their social nature. Without them the cohesion and stability of Roman society could not have been possible. It is significant that the Romans had no word for religion. 'Religio' simply meant 'matters of state'. The reason is very simple. There was no need for such a word, no more than there is in the case of any African tribe. All the beliefs and rituals which we regard as making up a society's religion were an essential part of its culture, which controlled its behaviour, i.e. which provided it with its effective government. As Fustel de Coulanges wrote, of the Ancient City: 'This State and its religion were so totally fused that it is impossible not only to imagine a conflict between them, but even to distinguish one from the other.'[6]

In the case of Rome, as in the case of many tribal societies, the land it occupied was itself closely associated with its religion. It was holy land, the land where the society's ancestors were buried. In the same way, Roman society was a holy society since its structure was sanctified by its gods whose own social structure closely reflected it.

It is not surprising that the Plebeians were originally excluded from active participation in public affairs. If they had no place in the body religious they could have none in the body politic with which it coincided.

The story of their mass departure from Rome, and voluntary exile to the Sacred Mountain is well known. They left saying: 'Since the Patricians wish to possess the City for themselves, let them do so at their leisure. For us Rome is nothing. We have neither hearth nor sacrifices nor fatherland. We are leaving but a foreign city. No hereditary religion attaches us to this site. All lands are the same to us.'[7] However, their voluntary exile was short-lived. This structureless mass of people was incapable of creating a city on the model of that which they had known. Consequently they returned to Rome and after many struggles established themselves as citizens of the Republic. If they were eventually enfranchised, it was that they had become culturally absorbed into Roman society, but for this to be possible the latter had to undergo considerable modifications. It could, in fact, no longer remain a tribal city.

Thus, whereas previously it was the Patricians, an aristocratic elite, who ruled Rome, a new elite slowly developed to replace it, the Senatorial class, composed of both Patricians and Plebeians. Its power was not based on hereditary status but much more on wealth. Such a change in itself must have seriously undermined the basis of social stability, by substituting the bonds established by 'contract' for those dependent on 'status' as a basis for social order. The resultant society, however, was still reasonably stable and probably would have lasted a very long time if it had not been for Rome's expansionist policies which led to the establishment of the Empire. The changes which this slowly brought about to every aspect of social life were far-reaching and profound and it is to them above all that one must look for the causes of the decline and fall of Rome.[8]

FOREIGN INFLUENCES

Foreign influences were undoubtedly the first cause of the changes which overcame Roman society. The cultural pattern which holds together the members in a traditional society and controls its relationship with its environment rarely survives the onslaught of powerful and unfamiliar foreign influences. Consider how that of the tribes of Africa has been disrupted by the colonial powers. The literature on the subject is voluminous. Think of the terrible cultural deterioration of the society of South Vietnam as a result of harbouring in its midst half a million affluent and pleasure-loving American soldiers. Look at the havoc at present being wrought to the very essence of Indian society by the cinema—which reflects a spirit totally alien to the Indian tradition.

But what were these foreign influences in Rome? First of all, after Sulla's conquest of Greece, Roman society was seriously affected by Greek influences. Greek literature, Greek philosophy, Greek manners, Greek dress became the rage. It spread from the fashionable circle of Aemilius Paulus and his friends to the people at large. These influences were not those of the Greek City States of the age of Pericles, but those of an already degenerate Greece—one that had been long subjected to auto-cratic Macedonian rule, that had largely forgotten its ancient traditions and with them its spirit of self-government. The warnings of the elder Cato were in vain. Greek influence was highly disruptive, just as several centuries later was the influ-ence of a decadent Rome on the German tribes that had the mis-fortune to be attracted within its orbit. (It is indeed no coincidence that such tribes as the Suevians, the Vandals, and the Ostrogoths, who figure so prominently in Roman history, disappeared without trace, while their more barbarous cousins, the Angles, the Saxons, the Jutes, the Helvetiae, and the Franks, etc., who succeeded in remaining outside the orbit of the Roman world, developed into the modern nations of northern Europe.)

To the Romans, however, more destructive than Greek in-fluence was that of the Eastern Provinces. In antiquity there had been stable traditional societies in Syria, Palestine, Persia and Mesopotamia. These however had been engulfed into the Empires of Babylonia, Assyria and of the Persian Achaemenids and their successors, and within them there had predictably arisen vast shapeless cities like Ninevah, Babylon, and Persepolis with their structureless and demoralized proletariats, which had much in common with the conurbations of our own industrial society.

Not surprisingly, there arose in these cities all sorts of monotheistic cults of the sort that tend to satisfy the requirements of alienated and demoralized people. As Roman society disintegrated, so did these cults appear ever more attractive to the urban masses of Rome. Indeed the vogue for Eastern religions spread throughout the Roman Empire. The philosopher, Themistius, during the reign of Valens writes of the 'mass and confusion of varying pagan religious views'.[9] He thought that there were at least 300 sects and 'as much as the deity desires to be glorified in diverse modes and is more respected, the less anyone knows about them'. Christianity was in fact but one of these importations. It is probable that if it had not been adopted by the Empire, another very similar one would have been adopted in its stead, possibly, as Ernest Renan suggests, Mithraism.[10]

If these cults spread so easily within the Roman population, it was because the ground was fertile for them. In fact the Roman people had grown more similar to those among whom these cults had originally evolved, and had developed the same psychological requirements which the cults were designed to satisfy. Undoubtedly they would have had little chance of spreading among the Romans of the days of Cincinnatus—no more indeed than would the strange sects (many of them too of Oriental origin) which are gaining ground today among the culturally deprived youths of our own conurbations to replace the discredited culture of industrialism.

OTHER-WORLDLINESS

I have already referred to the fact that the organization of the Roman gods reflected the highly structured Roman society of old. The Oriental cults which replaced the traditional Roman religion were very different. They were largely monotheistic and I think it can be shown that *monotheism is the creation of a structureless, and hence a disintegrated, society.*

In a stable tribal society, the supreme god plays but an accessory role, and is usually referred to as the creator or 'moulder'[11] and regarded as too divorced from human affairs to have any interest in tribal matters, let alone in those of the individual.

The society's protection is assured by the family gods and those of the clan and the tribe—ancestral figures who, rather than being regarded as having gravitated, at the time of their death, to some distant paradise, are, on the contrary, considered to have simply graduated to a more prestigious age-grade. Indeed, a tribe is often said to be composed of the living, the dead and the yet to be born—hence its great continuity or stability. The god of these new monotheistic cults, however, had no connection with any specific society, nor indeed any great interest in society as such. His interest was specifically with the individual. His character too had changed. He was no longer a worldly, pleasure-loving figure like Jupiter, but a stern autocrat, and this undoubtedly reflected the autocratic nature of government in the structureless society within which these cults originally sprang. Also he possessed a wife and child. This, too, was a great innovation. Their function was probably to satisfy psychological requirements previously fulfilled by the now defunct family gods, which the stern male god could not fulfil alone.[12]

It is interesting to note how the same changes overcame the pantheon of Ancient Egypt with the disintegration of Egyptian society during the Ptolemaic era. Indeed, with the triumph of Christianity, statues of Isis holding the baby Horus in her arms were frequently identified with the Virgin Mary and the baby Jesus, and unknowingly introduced into Christian churches.

As already pointed out, the essential feature of these cults was that they were eminently asocial. Duties towards society were replaced by duties towards God, and success in the next world was substituted, as the ultimate end of human activity, for success in this one. Lecky emphasizes this essential point throughout his history.

> The first idea which the phrase 'a very good man' would have suggested to an early Roman would have been that of a great and distinguished patriot, and the passion and interest of such a man in his country's cause were in direct proportion to his moral elevation. Christianity diverted moral enthusiasm into another channel and the civic virtues, in consequence, necessarily declined.[13]

Among other things, this meant that there was no longer any moral authority to prevent asocial behaviour. Thus one finds the extraordinary spectacle of people who succeeded in combining the pious fulfilment of their religious obligations with the most socially aberrant behaviour. As Lecky writes,

> the extinction of all public spirit, the base treachery and the corruption pervading every department of the government, the cowardice of the army, the despicable frivolity of character that led the people of Treves, when fresh from their burning city, to call for theatres and circuses, and the people of Roman Carthage to plunge wildly into the excitement of the chariot races on the very day when their city succumbed beneath the Vandal: *all these things co-existed with extraordinary displays of ascetic and missionary devotion*

As in our case, the inhabitants of Imperial Rome suffered from a cultural split-personality. They were denizens of two contradictory worlds—or rather half-worlds—since people have always required a spiritual as well as a physical existence—and in the stable societies of the past, they coincided.

THE IMPERIAL SYSTEM

People tend to ignore the important principle that autocracy is unknown among stable traditional societies. Yet it is a basic principle of stability that it can only be achieved by a self-regulating system.[14] When stable societies were governed by kings, these were in no way autocrats. The Homeric king like the original Roman king could be dethroned by a simple show of hands just as his West African counterpart can be 'destooled' in the same way.

In the Homeric city, the sovereign power resided not in the person of the king, in fact, but in public opinion: 'demouphemos', as it was referred to. This power was institutionalized into 'demoukratos', the latter without the former being but an empty facade. The role of public opinion was to ensure the maintenance of the traditional law.[15] This was further ensured by the prestige of the Council of Elders and also by the fear of incurring the wrath of the ancestral spirits—the ultimate guardians of tradition, and hence of the society's continuity or stability.

It is essential to realize that the kings of Rome were not thrown out because they were 'reactionary' and hence opposed to 'social progress'—the reason normally given for dethroning monarchs today—*but for precisely the opposite reason*. They were guilty of causing a departure from traditional law by attempting to incorporate the Plebeians into the body politic, which would have meant radically modifying its tribal nature.

In spite of the remarkable efforts of Augustus to establish a more 'flexible' type of government under his 'principate' which at the same time maintained as many features as possible of the Roman Republic, the Imperial system grew to resemble the latter less and less, until eventually it became its diametric opposite.[16]

The Caesars, as described by Suetonius in *The Lives of the Twelve Caesars* were Oriental tyrants with little regard for anything but the satisfaction of their ever more extravagant personal caprices.[17] Suetonius may have been biased, but it is also true that where there is smoke there is usually fire. The Antonines were undoubtedly an improvement, but after them the quality of the emperors went from bad to worse.

At the same time the influence of the Senate, which itself was undergoing a considerable change, waned increasingly. Eventually sovereignty resided, in effect, with the army and the urban masses. In the fourth century these took it in turns to name the emperor, whose average reign was little more than two years and who usually met his death by violence. During this period the increasingly degenerate Senate succeeded in naming but two emperors, both of whom were able to remain in power for but a very short period of time. In the fifth century and after—certainly after the reign of Theodosius—the emperors, as we have seen, were but puppets in the hands of some barbarian general or of some foreign power.

THE CHANGING ARMY

Roman expansion also led to radical changes in the nature of the army. During the early Republic, it was a citizen army. The soldiers were very much part of the body politic and its members owed allegiance to the Senate like all other Roman citizens. It was only after the Punic Wars, when Rome had acquired an Empire which stretched as far as Spain and North Africa, which meant that soldiers posted in these far away places were forced to remain for long periods away from home, that it became necessary to establish a professional army. As a result, soldiers became isolated from the rest of the body politic, and their loyalty to the Senate became progressively replaced by loyalty to their local commander. If it were not for this there would never have been the civil wars, nor for that matter would the Imperial system have taken the form it did in its later phases when the legions made and unmade emperors at their leisure.

Later, as we have seen, the commanders were increasingly foreigners, and eventually so were the troops themselves. This reduced still further the efficacy and loyalty of the Roman army.

PUBLIC GAMES

Another disintegrative influence was the institution of the public games. As Lecky writes, 'One of the first consequences of this taste was to render the people absolutely unfit for those tranquil and refined amusements which usually accompany civilisation. To men who were accustomed to witness the fierce vicissitudes of deadly combat, any spectacle that did not elicit the strongest excitement was insipid.'[18] Once more the parallel with our own disintegrating society is very striking. In order to pander to the increasingly barbarous audiences of twentieth century industrial society, TV and film producers vie with each other in their efforts to devise ever more garish spectacles based on increasingly exotic exploitations of the twin themes of sex and violence. Indeed not to do so, in the age we live in, would be to court inevitable bankruptcy.

It must be noted, however, that it is but in a very degenerate society that such spectacles would in the first place be tolerated let alone have any hope of flourishing. In tribal societies entertainments are of a very different kind. As Roy Rappoport shows, they take the form of feasts devoted to dancing, singing, eating and drinking in which everybody participates. These events are not frivolous entertainments. In all sorts of subtle ways, they play an essential role in maintaining a society's social structure and, at the same time, its stable relationship with its environment.[19] They are in fact rituals and as a society disintegrates, so do social activities become correspondingly deritualized. As they do so they lose their social purpose, and become, socially speaking, random events: cathartic outlets at best.

SLAVERY

The institution of slavery was undoubtedly another cause of the fall of Rome. Among other things, it permitted the growth of large-scale business enterprises with which neither the artisan nor the small farmer, who were proverbially the backbone of the Roman Republic, could possibly compete. As was the case with our own Industrial Revolution, and as is today the case in the Third World with the advent of the Green Revolution, artisans and small farmers were ruined and inevitably migrated to the cities to swell the ranks of the depressed urban proletariat. Rome had increasing difficulty in accommodating them. They were made to live in squalid and overcrowded conditions and to provide them with their sustenance became one of the major preoccupations of the Roman State, causing it to indulge in ever more distant military campaigns for booty and tribute and to adopt ever more destructive agricultural methods.

But the institution of slavery had other consequences too. The great wealth which it generated changed, perhaps more than anything else, the nature of Roman society. It provided a striking contrast with the material austerity which was an essential feature of early Roman society and which was possibly a condition for the maintenance of the other virtues which characterized the Romans during that period.

It is worth quoting Lecky's description of the situation which developed in the capital as a result of these disastrous trends.

> The poor citizen found almost all the spheres in which an honourable livelihood might be obtained wholly or at least in a very great degree preoccupied by slaves, while he had learnt to regard trade with an invincible repugnance. Hence followed the immense increase of corrupt and corrupting professions, as actors, pantomimes, hired gladiators, political spies, ministers to passion, astrologers, religious charlatans, pseudo-philosophers, which gave the free classes a precarious and occasional subsistence, and hence, too, the gigantic dimensions of the system of clientage. Every rich man was surrounded by a train of dependants, who lived in a great measure at his expense, and spent their lives in ministering to his passions and flattering his vanity. And, above all, the public distribution of corn, and occasionally of money, was carried on to such an extent, that, so far as the first necessaries of life were concerned, the whole poor free population of Rome was supported gratuitously by the Government.[20]

THE FREE DISTRIBUTION OF CORN

This played the same role as does our own state welfare system in the decaying conurbations of the industrial world. In New

York well over a million people (1975) subsist entirely on state welfare. In the US, as a whole, the figure is about fifteen million (1975). In Britain, according to a study undertaken by the Department of Health and Social Security in the 1970s, within the next two decades social work and 'caring' organizations will be one of the most significant employers in the country. In Rome, the scale of the free distributions of corn was eventually such that, in the words of Lecky,

> To effect this distribution promptly and lavishly was the main object of the Imperial policy, and its consequences were worse than could have resulted from the most extravagant poor-laws or the most excessive charity. The mass of the people were supported in absolute idleness by corn, which was given without any reference to desert, and was received, not as a favour, but as a right, while gratuitous public amusements still further diverted them from labour.

The greatest damage done by state welfare, however, is to bring about the disintegration of the family unit itself. Indeed, this basic unit of human behaviour, without which there can be no stable society, cannot survive a situation in which the functions which it should normally fulfil have been usurped by the state. The family, in traditional societies, is an economic unit, as well as a biological and social one. If the father and the mother no longer have to make any effort to feed their children, if they no longer have to ensure their proper upbringing and education, then it must almost certainly decay.

A society in which the family has broken down is in the final stages of disintegration. Such is the case today in the ghettos of the larger American conurbations—not only in the US but in Mexico, in Venezuela and elsewhere. Such a society is characterized by all possible social aberrations, such as crime, delinquency, vandalism, drugs, alcohol, etc. which are indulged in by people as a means of divorcing themselves as much as possible from a social environment which is increasingly intolerable. One can only assume that these social deviations also characterized the depressed areas of urban Rome during the later Empire.

THE CONSEQUENT BREAKDOWN

All these changes led to the total demoralization of the Roman people and to the elimination of those qualities to which must be attributed the success of the Roman State in its earlier phases. One can do no better than quote Lecky again on this subject,

> All the Roman virtues were corroded or perverted by advancing civilisation. The domestic and local religion lost its ascendancy amid the increase of scepticism and the invasion of a crowd of foreign superstitions. The simplicity of manners, which sumptuary laws and the institution of the censorship had long maintained, was replaced by the extravagancies of a Babylonian luxury. The aristocratic dignity perished with the privileges on which it reposed. The patriotic energy and enthusiasm died away in a universal empire which embraced all varieties of language, custom and nationality.[20]

Lecky could not be more eloquent in his description of the degeneration of Roman morals especially when he compares them with the moral qualities of the Romans during the early Republic.

> In the Republic when Marius threw open the houses of those he had prescribed to be plundered, the people, by a noble abstinence rebuked the act, for no Roman could be found to avail himself of the permission. In the Empire, when the armies of Vitellius and Vespasian were disputing the possession of the City, the degenerate Romans gathered with delight to the spectacle, as to a gladiatorial show, plundered the deserted houses, encouraged either army by their reckless plaudits, dragged out the fugitives to be slain, and converted into a festival the calamity of their country. The degradation of the national character was permanent. Neither the teaching of the Stoics, nor the government of the Antonines, nor the triumph of Christianity could restore it.

SOIL DETERIORATION

At once both a cause and an effect of social breakdown was the steady decline in the productivity of the land from which the Roman masses drew their sustenance.

Like all governments which depend for their survival on the support of a growing urban population, that of Rome adopted a cheap food policy. Just as in Britain, agricultural decline was partly caused by the import of cheap foods from abroad whenever the occasion arose. It was also caused by forcing the farmers to pay taxes which they could not afford, and also, as we have seen, by creating conditions in which only the largest enterprises could survive. Everything was geared to the short-term and, just as with us, the long-term consequences of the totally unsound agricultural practices which inevitably arose led to the most terrible deterioration of the soil. To this day, Southern Italy is a semi-wilderness, and the Italian Government, rather than attempt to restore the soil's lost fertility, is misguidedly attempting to combat the resultant poverty with large-scale industrialization.

The deserts of North Africa, which was once the granary of Rome, bear even more eloquent testimony to the destructive agricultural practices of the times. The area bristles with the ruins of once magnificent cities. Thus, where there is now the wretched village of El Jem, once stood the Roman city of Thysdrus, of which the most conspicuous remnant are the ruins of a colosseum which once seated 65,000 spectators. Where there is now the equally wretched village of Timgad, once stood the great city of Thamugadi, built by Trajan in the year AD 100. That city was once supported by extensive grain fields and olive orchards, of which there are now no trace. To quote Carter and Dale

Water erosion, as well as wind erosion, has been at work on the landscape. Gullies have been cut out through portions of the city and have exposed the aquaduct which supplied the city with water from a great spring some three miles away. Ruins of the land are as impressive today as the ruins of the city. The hills have been swept bare of soil, a story which may be read throughout the region.[21]

Those who refuse to face the terrible destructiveness of large-scale commercial farming have often argued that all this was caused by a change in the weather. Research however shows that this was simply not so. Evidence of unchanged climate in the last two thousand years is provided by the successful planting of olive groves on the sites of ruins of Roman stone olive presses.

Let us look a little more closely at the development of Roman agricultural practices. The Etruscans, whose influence spread throughout most of Italy, before it was halted by the power of Rome, appear to have been remarkable farmers. Their underground drainage systems were masterpieces of engineering, which nobody today could possibly afford to build. The Romans, therefore, appear to have inherited reasonably fertile land. To begin with, the Republic occupied but a small area of Latium, perhaps no more than 400 square miles, but its population was very dense: apparently, something like a thousand persons per square mile. It is suggested by Carter and Dale that if the Romans were mainly vegetarians it may well have been because they had to, since it needs a great deal of land to provide people with a meat diet. As could be expected, the average farm holding was very modest: between one and five acres of arable land. The farmers produced mainly for themselves, though a small surplus was sent to Rome to feed the relatively small urban population.

It may well be that land shortage was one of the principal reasons for Rome's initial policy of expansion. In any case, it was this policy which appears to have caused the first major ecological catastrophe for, among other things, it meant cutting down forests for the timber needed to build the requisite fleet. Thus, when England built her fleet to fight the Spanish Armada, not only was it necessary to cut down all suitable trees at home, but more had to be imported for this purpose from as far afield as Norway. In the same way, surrounding mountainsides and their foothills were stripped of their forest cover to supply timber for the Roman fleet during the First Punic War. At the same time, much farmland had to be abandoned because the farmers were serving in the army, and later because of the devastation caused by Hannibal and his troops in Italy.

Significantly, it was around 200 BC that malaria seems to have developed in Italy. The marshes which bred the Anopheles mosquito were man-made. They were created by soil erosion from badly-cultivated sloping lands, which, being deprived of their topsoil, were, by the same token, made largely unproductive. The Pontine marshes supported sixteen Volscian towns in the seventh century BC. Five hundred years later, they only supported mosquitoes.

Also significantly, it was after the Second Punic War that Rome first became a grain importing country, or more precisely, that grain had to be obtained in the form of tribute from Sicily, Sardinia, Spain and later North Africa, to feed the urban population.

It was also during the first and second centuries BC that the latifundia began to replace small farms; at the same time arable land was converted wholesale to pasture land, while much of it was abandoned since its soil had become totally exhausted.

There is a considerable literature on this subject; Lucretius for instance described it in detail.[22] He believed that the earth was dying. Livy wondered how the vast armies of the Volscians, Aequians and Hernicians, which the Romans had fought four centuries earlier, could have been sustained by lands which, in his time, were so poor that they could support but a small population of slaves tending their masters' livestock and caring for sparse olive groves and vineyards.[23] St Cyprian of Carthage, around AD 250, complained that the world was dying. Springs were drying up and famines were increasing over the whole Mediterranean area.[24]

Most eloquent perhaps was Columella, who argued against the thesis that the deterioration of the land was a natural process—'an act of God', as we would refer to it—the consequence of the earth's natural ageing process.

> It is a sin, to suppose that nature, endowed with perennial fertility by the creator of the universe, is affected with barrenness as though with some disease, and it is unbecoming to a man with good judgement to believe that the earth, to whose lot was assigned a divine and everlasting youth, and who is called the common mother of all things—because she had always brought forth all things and was destined to bring them forth continuously—has grown old in mortal fashion. Furthermore, I do not believe that such misfortunes come upon us as a result of the fury of the elements, but rather because of our own fault.[25]

He attributed them to the disappearance of the yeoman farmer and his replacement by slave labour. Slaves simply did not bestow upon the soil the care it required—also, the owners of the latifundia, in their haste to get quick returns, starved the soil of manure. For all land exhausted by cropping, he writes, 'there is one remedy, namely to come to its aid with manure.'

Columella describes how farmers made use of different types of manure, bird manure, human waste solid and liquid, animal manure. Their discriminating use by Roman farmers reflects generations of experience of careful small-scale farming designed as much to produce high quality food as to preserve the soil's food-producing capacity. Columella considers that it is a lazy farmer who cannot get at least 20 bushels per month from his small animals. He also dwells on the practice of interchanging soils, which was also done in traditional Chinese farming. This involves bringing in or taking out, with an extraordinary expenditure of labour, different soils, in hand baskets, to provide for the requirements of different crops. To the Roman farmer this laborious process was known as 'congestica'. As Columella wrote, 'it is not because of weariness as many believed, nor because of old age, but manifestly because of our lack of energy that our cultivated lands yield us a less generous return.'[25]

In our own era, a succession of wise and learned agricultural theorists such as Sir Albert Howard, Lord Lymington, Sir George Stapleton, and Robert Waller, have warned succeeding British governments of the folly of our short-term agricultural methods, and shown that no sustainable agriculture is possible without systematically returning organic matter to the soil.

All this drastically changed once large-scale agriculture replaced subsistence farming, and the produce of the land was transported to distant markets to be consumed by a predominantly urban population. It also changed as the daily collection of night soil to fertilize neighbouring fields and gardens was abandoned and elaborate sewage systems installed. It is indeed proverbial that the fertility of North Africa went down the sewers of Rome. Sadly, we have learnt little from this experience and today are repeating the same unpardonable error on a vastly greater scale.

VULNERABILITY

As Rome became increasingly dependent on the empire for its sustenance, so did life within the city become increasingly precarious. Instability is a necessary feature of a system which expands anarchically beyond its optimum size.

Tacitus observed as early as the reign of Claudius that Italy, which had once supplied the distant provinces with corn, had become dependent for the very necessities of life upon the winds and the waves.[26] The situation, as Lecky points out, was hopeless, 'Adverse winds, or any other accidental interruption of the convoys of corn, occasioned severe distress in the capital; but the prospect of the calamities that would ensue if any misfortune detached the great corn growing countries from the Empire, might well have appalled the politician'.

This nearly occurred with the revolt of Gildo, in Africa, in AD 397-8, and actually did occur, when, in the fifth century, the kingdom of the Vandals established itself at Carthage and grew to become a great naval power, its fleet gaining control of the Mediterranean itself. As a result, it lay within its power to cut off Rome from its food supplies in North Africa, just as it is in the power of the Arab Sheiks today to cut off the industrial world from their only available source of cheap energy on which their economy, indeed their very sustenance, entirely depends.

However the Romans, by this time, had become too weak and too demoralized to oppose the Vandals with the force of arms, and when the Vandal King, Genseric, advanced with his army on Rome in AD 470 there was no army there to oppose him. All the degenerate Romans could despatch against him was a procession led by priests, pathetically appealing to him to spare their city from the flames.

Rome never really recovered from this blow, and when the Vandal Kingdom was eventually destroyed by Justinian, its fate was already sealed.

FINANCIAL PROBLEMS

A further problem was the financial cost of providing free corn and free games for the multitudes and that of paying for the army which was constantly clamouring for more money, and whose power grew in direct proportion to the helplessness of the people, resulting from the breakdown of their social structures. Failure to satisfy its demands generally led to the overthrow of the emperor and his replacement by one more sympathetic to army interests.

As social and ecological disruption intensified, the Government was faced with the increasingly serious financial costs of attempting, as it did, to mask the symptoms of a disease which it could not cure. Thus, to raise money in the form of tributes from conquered territories became increasingly difficult. Settled populations around the Mediterranean had already been conquered and expeditions had to be undertaken into ever more remote areas. The point was eventually reached when there were no peoples left to conquer who were worth conquering. A possible exception were the Persians, but those indomitable people the Romans could never subdue.*

One means of raising money, needless to say, was to increase taxes, and they eventually became prohibitive. Indeed if the rural population started grouping itself around local strong leaders, thereby giving rise to the feudal system, it was not in order to obtain protection against marauding bands, but against government tax collectors (Curiales).[27]

In such conditions, the State's financial situation grew steadily worse and, not surprisingly, it resorted to precisely the same expedients to which governments of today are resorting to tide over their immediate financial problems regardless of the social and ecological costs involved. Thus we find debased currency

---

*In 53 BC the army commanded by the Consul Crassus was annihilated by the Parthians. In AD 260 the Roman Emperor Valerian was captured, together with his army. Little more than a century later the Emperor Julian was killed when leading an expedition against the Sassanids who also defeated his successor, the Emperor Jovian.

being put into use; silver and gold coins which contained neither silver nor gold. Inflation, needless to say, grew eventually out of control.

GOVERNMENT REACTION

The Roman politicians, like ours today, never really understood the problems with which they were faced. The Gracchi thought they could solve the problem by legislation. Land distribution under the Gracchi law, was a failure. There is no point giving people a few acres of worn-out land if social and economic conditions do not favour the survival of the small farm. *It is the social and economic conditions which must be changed.*

Once the fertility of the soil had become so reduced that some of the latifundia did not even produce enough grain to feed the slaves who worked them, the Emperor Domitian issued an edict forbidding the planting of grapes in Italy. He even went so far as to order each landowner in the provinces outside Italy, to destroy half the grape vines. Needless to say, that edict was un- enforcible and had to be repealed. If grain was not grown, it was because socio-economic conditions did not permit it. *It was those conditions which had to be changed,* and *this he was powerless to do,* just as our politicians are powerless to change those in which we live, and which render unfeasible any sustainable agricultural practices.

Pertinax, in AD 193, offered to give land to anyone who would cultivate it, but there were few takers.[28] Eventually, Diocletion, AD 284-305, attempted more drastic measures and issued an edict binding all free farmers and slaves to the land which they occupied. This was the beginning of the 'Coloni' system which eventually led to mediaeval serfdom.[29]

It is interesting to note that in 1975 both in China and the Soviet Union, the intolerable trend towards urbanization was also halted by legislation. The peasants were simply not allowed to leave their villages and were effectively tied to the land. These so-called progressive states have thereby adopted one of the principal features of the mediaeval feudalism which they have so vehemently decried.

However, such legislation was, in the chaos of the disintegrating Roman Empire, very difficult to apply, and in effect no government was ever able to contribute significantly to the reversal of the process of social disintegration, to which the Empire was irreversibly condemned. The wise Emperor Marcus Aurelius advocated a stoical attitude towards a process which could not be halted. As Lecky points out, no emperor could have survived an effort to eliminate the more obvious causes of social and ecological degradation. The free distribution of corn had become an essential feature of life in the capital, to which everything else had adjusted. The people were addicted to it, just as the population of our larger conurbations is becoming addicted to our state welfare system.

Against slavery they were equally powerless. To interfere with it would have been to disrupt the economy itself, which was as dependent on slavery as ours is on machinery. Without slaves, Rome's already precarious agricultural system would have collapsed, just as would today our equally precarious one were it to be deprived of its tractors and chemical poisons.

To re-establish a socially orientated religion in a mass society which had been deprived of its basic social structure, and which had largely forgotten its ancient traditions, was an equally hopeless task. The Emperor Julian tried. For his efforts, he has become known to history as the Apostate.

To reform the political institutions was equally impossible. Augustus, after the civil wars, tried to do so, or rather to reconcile them with his Principate. His effort was ingenious but the structure he built was a bastard one. The institutions of the Roman Republic were admirably suited to a City State but not, as Hammond shows, to an Empire.[30] It is not surprising that such diametrically opposite social forms could not be forced into the same institutional framework. By becoming an Empire, Rome was in fact forced to break away from its past, and to do so, as Burke pointed out, 'is the greatest tragedy that can befall a nation.'

Diocletian's efforts at reform were even less successful. In the absence of a real society held together by public opinion reflecting its traditional cultural pattern, there was no power-base other than the army and the urban masses, both of which were only concerned with the acquisition of increasingly short-term benefits. This is true today of most of the disintegrating societies throughout the world, as they fall within the orbit of the industrial system.

In order to combat inflation, Diocletian tried to break the monopolies and suppress combinations in restraint of trade by fixing maximum prices for principal commodities such as beef, grain, eggs, clothing; in all seven to eight hundred items. In this respect, he was more thorough than our politicians, for the death penalty was prescribed for anyone who sold these goods at a higher price. He also fixed the wages of teachers, advocates, bricklayers, weavers, physicians and even of various unskilled labourers. Needless to say, efforts to control prices and wages proved as ineffective as they have in modern times.

The situation could only go from bad to worse. By systematically accommodating undesirable trends rather than reversing them, a chain reaction was set in motion which could only end in total collapse. Thus, welfare caused the increasing demoralization of the urban masses. The public games made them still more degenerate. Both helped attract more people to the cities, thereby increasing the dimensions of the problem.

In the meantime, the inevitable depopulation of the country-side and the ruin of the yeomen and rural middle classes, made agriculture increasingly dependent on slave labour, further accelerating rural depopulation and further swelling the urban masses. The shortage of money to buy food and the inevitable soil deterioration required further taxation and ever more expeditions undertaken in search of tribute and booty by an ever less effective army, whose loyalty to the State was ever more in doubt, and which eventually simply disintegrated along with the rest of the body politic.

Our politicians are today caught up in a very similar positive feed-back process, from which they appear even less capable of extracting us. With neither vision nor courage, they simply allow the Ship of State to drift into ever more turbulent waters and content themselves with superficially repairing its ever more battered hulk, for no other purpose than to defer, for ever shorter periods, the inevitable day when it must flounder beneath the waves. Such is the price which must be paid if social and ecological exigencies are subordinated to short-term political and economic interests.

Notes:

1. Edward Gibbon, *The Decline and Fall of the Roman Empire*, London, Alex Murray & Co, 1870.
2. Samuel Dill, *Roman Society in the Last Century of the Western Empire*, New York, Meridien Books, 1958.
3. Edward Gibbon, op.cit.
4. Leon Homo, *Les Institutions Politiques Romaines*, Paris, La Renaissance du Livre, 1927.
5. Ennius, *Annales* Quoted by Hammond, *City-state and World State in Greek and Roman Political Theory until Augustus*, Harvard University Press, 1951.
6. Fustel de Coulanges, *La Cite Antique*, Paris, Hachette, 1927.
7. W. W. Fowler, *The City State of the Greeks and Romans*, London, Macmillan, 1921.
8. Sir Henry Maine, *Ancient Law*, London, 1861.
9. Themistius, Fragments, quoted by Jacob Burckhardt in *The Age of Constantine the Great*, London, Routledge and Kegan Paul, 1949.
10. Ernest Renan, *Marc Aurele et la Fin du Monde Antique*, Paris, 1882.
11. Edwin W. Smith, editor, *African Ideas of God*, London, Edinburgh House Press, 1950.
12. Eric Fromm, *The Art of Loving*, Unwin, London, 1957.
13. Edward Hartpole Lecky, *History of European Morals from Augustus to Charlemagne*, London, Longmans, Green & Co., 1905.
14. Edward Goldsmith, A Model of Behaviour, *The Ecologist*, Vol.2, No.12, December 1972.
15. G. Glotz, *The Greek City and its institutions*, London, Routledge & Kegan Paul, 1921.
16. Mason Hammond, *City-state and World State in Greek and Roman Political Theory until Augustus*, Harvard University Press, 1951.
17. Suetonius, *The Lives of the Twelve Caesars*, New York, Random House, 1931.
18. Edward Hartpole Lecky, op.cit.
19. Roy A. Rappoport, *Pigs for Ancestors*, New Haven, Yale University Press, 1967.
20. Edward Hartpole Lecky, op.cit.
21. Vernon Gill Carter and Tom Dale, *Topsoil and History*, The University of Oklahoma Press, 1974.
22. Lucretius, *De Rerum Natura*, Penguin Books, London.
23. Livy, Quoted by Carter and Dale, op.cit.
24. St Cyprian, Quoted by Carter and Dale, op.cit.
25. Columella, *On Husbandry (De Rerum Rustica)*, London, 1745.
26. Tacitus, Quoted by Carter and Dale, op.cit.

27. W. E. Heitland, *Agricola—A Study of Agriculture and Rustic Life in the Greco—Roman World from the point of view of Labour*, Cambridge Univ. Press, 1921.

28. Pertinax, Quoted by Carter & Dale op.cit.

29. Diocletian, Quoted by Carter & Dale, op.cit.

30. Mason Hammond, op.cit.

CHAPTER TWO

Education:
What For?

' . . . he's a compulsive buyer, thrives on additives and industrial waste, doesn't need a home or a family, has no sense of aesthetics and, in our high-technology schools, gentlemen, we can churn him out by the thousands.'

Education:
What For?

FEW PEOPLE TODAY would dispute that education is a good thing. Most would even consider that the more we get of it the better. In fact it is increasingly regarded as an inalienable right of all citizens, regardless of ability. The reason is that we believe it to be the key to success in the industrial world we live in, as is 'manna' among the Polynesians, 'muntu' among the East Africans, 'baraka' among the Arabs—a sort of vital force on whose accumulation success in life ultimately depends. As a result, we spend an ever increasing proportion of the national budget on education, and an ever increasing number of our youth are made to spend an ever greater part of their lives in educational institutions. What is the result of these efforts?

Literacy, contrary to what one would expect, is decreasing.[1] According to the British Association of Settlements there were two million illiterates in the UK in 1971 and the preponderance of illiterate adults rather than belonging to the older generation as one would expect, were aged twenty-five and under.[2] A report some years ago *Trends of Reading Standards* confirms what the late Sir Cyril Burt wrote in the 'Black Papers' on education, that standards of literacy are today lower than they were in 1914. What appears extraordinary is that literacy seems to have been going down fairly steadily ever since the state took an active part in education.

At a global level, the situation appears even worse. A 1972 UNESCO study of world education, reports that the number of illiterates over fifteen years of age increased between 1960 and 1970 from 735 million to 783 million, though admittedly the proportion of illiterates over fifteen dropped from 44.3 per cent to 34.2 per cent of the adult population. The number of children dropping out of school without taking any examinations remains remarkably high.[3]

In the north of England, Yorkshire and Humberside, the figure was over 50 per cent. In much of the Third World, according to a UNESCO report, drop outs from primary education are as much as 80 per cent of those who enrol, and this is in countries in which only 10 per cent of children between the ages of six and twelve have attended school. This means that only 2 per cent of children pursue their studies to the end of the primary school programme.[4] These trends have continued since this chapter was written in 1973.

DISCIPLINE

Another change is that schools have become far more permissive. Teachers no longer command the obedience they used to. In many schools, especially in the slums of the larger industrial cities, it is increasingly difficult for them to keep order and often attempting to do so occupies so much of their time that little is left for teaching. In many cases teachers are abused and even assaulted by the pupils.

THE PRACTICAL RESULTS OF EDUCATION

Although our educational system provides many of the skills required for the functioning of our industrial society, success for those who pass their exams is not necessarily assured. Unemployment among school leavers is high, as it is among university graduates. In universities in particular, the courses do not necessarily relate to the demand for specific skills. In the USSR, as in India and other parts of the Third World, there is a surplus of engineers; in France of lawyers; in the US of physicists; while throughout the industrial world there is a growing shortage of craftsmen, such as carpenters and plumbers—the products of a very different type of educational system.

As expenditure on education is increasing much faster than gross domestic product (GDP), the economy's capacity to absorb graduates is likely to continue declining and, as a result, the aspirations of only an increasingly smaller section of school leavers can be satisfied. The vast majority of graduates, condemned to fulfilling functions they have been taught to regard as menial, may be faced with a miserable and frustrating existence in jobs requiring skills for which they have had no specific training and for which their education has rendered them psychologically unfit.

This has already happened in the UK, where people often simply refuse to fulfil what they regard as low-prestige jobs.In the past, this has meant that to fulfil these functions society has taken in people from foreign lands whose education has not imbued them with the same set of prejudices. Thus in the UK we import the waiters in our restaurants from Italy, Spain and Cyprus, domestic servants from Portugal and the Philippines, workers in the construction industry from Jamaica, and bus drivers from the Punjab.

In this way, as irony would have it, the furore for mass education is leading, among other things, to the creation of a caste system—the proverbial epitome of social inequality.

In the meantime this massive educational effort is not making our society a visibly better place to live in. It seems that we face more crime, more delinquency, more alcoholism, more drug addiction and more of all the other problems associated with a disintegrating society.

COSTS OF EDUCATION

Even if modern education provides the benefit it is supposed to, it is a luxury that few countries can afford. In the UK it is increasingly obvious that we cannot possibly afford to increase expenditure on education. Indeed, if we project current trends of the GDP and current educational costs, we find that even if the whole of the government budget—which we can take as roughly 40 per cent of GDP—were to be devoted to education, we would still fail by about the year 2007 to meet foreseeable educational costs.

If the US were to introduce what educators call 'Equal treatment for all' in all state schools, the cost, according to Ivan Illich, would be somewhere around 80 billion dollars.[5]

In Britain we already spend on education twice the total income of the average Indian or Nigerian. In spite of this everything is being done by governments and international bodies such as UNESCO to spread western education throughout the countries of the Third World in the full knowledge that none of them can remotely afford it.

In Britain it seems that the point has already been reached when it will only be possible to improve our educational system by methods which do not require further investments. This rules out further centralization and further increases in the capital intensity of education. What it implies, in fact, is a complete change in our philosophy of education.

WHAT IS EDUCATION?

In spite of the extraordinary importance we seem to attach to education, nobody has really considered what it is for, nor why we in fact need it at all. Yet it must be clear that unless we can answer these questions we are unlikely, except by chance, to devise a satisfactory educational policy. Now to understand education, like all other human activities, one must look at it in a far wider context than we are accustomed to. When we talk of education we invariably mean western education. It occurs to few people that every one of the thousands of traditional societies studied by anthropologists has also developed its own educational system, often a very elaborate one at that.

Still less do we look at education in non-human animal families and societies; yet in many animal species a considerable amount of information must be communicated from one generation to the next via the family and sometimes the society.

It is known, for instance, that the larger predators, such as the lion and the tiger, must learn to hunt. Those brought up in a zoo would be almost certainly incapable of surviving in the challenging conditions of their natural habitat. Even apparently more modest animal species must learn—the chaffinch for instance cannot sing unless it is taught to do so. In this chapter I shall try to consider education in its widest possible context.

What then is education? Margaret Mead defines it as 'the cultural process . . . the way in which each new born individual is transformed into a full member of a specific human society, sharing with the other members a specific human culture.'[6] It is in fact but another word for socialization. It transforms an unspecialized child born with the potential for becoming a specialized member of a very large number of different social systems into a specialized member of a specific social system.

In a still wider behavioural context one can compare a child in a society with a cell in a biological organism. Immediately after division the latter is in possession of the full complement of hereditary material and is thereby capable of a very wide range of responses. Slowly, however, it becomes specialized in fulfilling that narrow range of responses required of a differentiated part of a biological organism. If we accept this definition, then the education implications are considerable. For instance, the Freudian notion of the community and the family as frustrating and as the cause of psychological maladjustments must be totally rejected. The very opposite appears to be the case, psychological maladjustments for the most part being the result of social deprivation.

It also means that so-called progressive education, in which parents and teachers allow children to do precisely what they like for fear of 'frustrating' them, is totally misguided. It is only by subjecting a child to a specific set of constraints, which it is probably only too happy to abide by, that it becomes capable of fulfilling its specific functions within its family unit and later within its community, i.e. that it can become socialized or, in fact, educated. Otherwise it remains isolated, goalless and alienated: as is increasingly the fate of much of our youth today.

Permissiveness at least with regard to matters of social significance is not a feature of stable social systems. On the contrary, their members tend to be disciplined. Discipline, in fact, appears to be a *sine qua non* of self-government. This discipline is displayed naturally. It is the product of socialization and of public opinion.

Significantly, to the Greeks liberty did not mean permissiveness but rather self-government. They were free, not because they were permissive, but because they were in charge of their own destinies while the Persians were slaves because they were ruled by an autocrat.

LEARNING AS A NORMAL PROCESS

If education is a behavioural process then it is subject to the laws governing other such processes. One such law is that behaviour proceeds from the general to the particular.[7] For this reason it is the earliest phases of education which are the most important. It is during those phases that the generalities of a child's behaviour pattern will be determined, while during the later phases they will simply be differentiated so as best to permit their adaptation to varying environmental requirements.

It must follow that the mother is the most important educator, and the quality of the family environment the most significant factor in determining a child's character and capabilities.

SEQUENTIAL EDUCATION

Another such law is that behavioural processes are sequential. Their various stages must occur in a specific order. If one is left out, then the subsequent ones will either not be able to occur at all, or will occur at best imperfectly. Thus, what a child learns during its formal institutionalized education cannot make up for any deficiency in the earlier phases of its upbringing within its family. This is the conclusion that most serious studies have revealed. J.S. Coleman, for instance, whose massive study *The Adolescent Society* led him to examine the career of 600,000 children, 6,000 teachers and 4,000 schools, reported in 1968 'that family background differences account for much more variation in achievement than do school differences'.[8] This is also the conclusion of the US Government study, *Equality of Educational Opportunities*, published in 1964 which stated that, 'variations in the facilities and curriculum of the schools account for relatively little variation in pupil achievement . . .' The most important factor measured in the survey is the home background of the individual child. In fact, whatever the combination of non-school factors, 'poverty, community attitudes . . . which put minority children at a disadvantage in verbal and non-verbal skills when they enter the first grade, the fact is the schools have not overcome it.'

One of the greatest problems that teachers have to face today is the proliferation of so called emotionally unstable children. These are exceedingly difficult to teach as they are unruly, un-disciplined and unable to concentrate on anything that is not ob-viously relevant to the satisfaction of their most short-term re-quirements. These are the children which are the most likely to become delinquents, criminals, drug addicts: the ones that cannot be socialized because the first phases in the socialization process, those which should have occurred in the home, were so deficient.

In a society in which the family unit has broken down, and whose principal institutions conspire to cause its further disinte-gration, the problem cannot be solved. Characteristically we choose to ignore the pre-school stages of education and reserve the very term for that which occurs at school. In this way, we define the educational problem in precisely that way which makes it appear amenable to the only sort of solution which our society can provide: the building of more and bigger schools, filled with ever more expensive equipment—language labor-atories, computers, tape recorders and God knows what—into which we consign our children for an ever greater proportion of their lives. Every government in turn contributes piously and self-righteously to this fatal process.

The present trend is to raise the school-leaving age, and the number of nursery schools, and crêches for working mothers, accommodating in this way the trend towards the further disintegration of the family. In reality, the only possible way to solve the problem is to reverse such trends, and hence to restore the family to its educative role which, in traditional societies, it has always enjoyed.

SOCIAL STABILITY

Another basic feature of all behavioural processes is that they tend towards stability. Stability is best regarded as a state in which a system can preserve its basic structure in the face of change. It is in effect but another word for 'survival' taken in its widest sense. In a stable system, discontinuities will be reduced to a minimum. This is only possible if environmental changes occur within certain limits. If they are too radical or too rapid, natural systems have no means of adapting to them.

The behaviour of human societies is in no way exempt from this rule, yet, in our industrial society, we set out purposefully to defy it. We tend to regard everything conducive to change as desirable. Our educational system puts a premium on innovation and originality in all its forms, i.e. it is geared to instability rather than stability.

In a traditional society, the opposite is the case. The basic preoccupation of its citizens is to observe the traditional law and to divert as little as possible from the cultural norm. Everything conspires to this end, since all deviations are seriously frowned upon by public opinion, proscribed by the council of elders and, it is believed, punished by the ancestral spirits. Education in such societies, as Margaret Mead writes, 'is the process by which continuity was maintained between parents and children, even if the actual teacher was not a parent but a maternal uncle or "shaman" '.

When a society becomes unstable, when social control breaks down and discontinuities grow ever bigger, then it is but a question of time before it eventually collapses. It is towards such a collapse that our educational system, together with the rest of the institutions of our industrial society, are leading us. To avoid it, education must, among other things, be designed to promote stability rather than change—but this cannot be done in an industrial society in which the promotion of instability, implicit as it is in our notion of progress, is the avowed object of public policy.

INFORMATIONAL FEEDBACK

Another feature of behavioural processes is that they involve feedback. Systems can only adapt to their environment because they are linked to it by means of all sorts of different feedback loops. If these loops are severed, as occurs once social behaviour becomes 'institutionalized' then they become isolated, can no longer satisfy environmental requirements and, from the point of view of the larger system, become random.

The introduction of random information into the system from the outside must have similar effects. Such random information will affect the generalities of a child's learning process, which will colour the subsequently developed particularities of its world-view. As our society 'progresses', so are its children bombarded with ever greater quantities of random information. Obvious sources are television personalities, newspapers and, unfortunately, one must include to an ever greater degree our educational system itself which is ever more isolated from the social process.

A living system will only tend to detect and interpret signals (and hence acquire information) that are relevant to its behaviour pattern, filtering out, so to speak, those that are irrelevant to it. Adult humans can undoubtedly do this, but for a child it is more difficult. Its behaviour pattern is still embryonic as must be the process of filtering out irrelevant signals. This essential function is thereby assumed by the child's family and community, from which relevant information must largely be derived, if the child is to be properly socialized and hence educated. However if these essential social groupings have disintegrated, then the child is helpless in the face of disruptive signals, and the information it builds up can only serve to mediate aberrant behaviour.

From the educational point of view, the implications of this principle are enormous. At the present time it is generally accepted that knowledge is good and the more the better. In reality only relevant knowledge is good, and then only if it is communicated in the correct sequence. Most of the knowledge we impart to our children is educatively 'random' and must actually impede rather than favour the process of socialization.

If one wants to be a purist, one can go so far as to say that the invention of writing was actually a blow to the cause of social stability. Quite apart from causing a drastic polarization of society between the literate and the non-literate, it also provided a store of information which entered into competition with traditional cultural information hitherto transmitted orally from one generation to the next. In this respect it is very much as if a DNA data-bank were set up to assist in the transmission of genetic information from one generation to the next. However, it seems unlikely that we shall ever unlearn it. Nevertheless, if we wish to reconstitute a society with any semblance of stability, steps will have to be taken to bring the current explosion in the mass media, particularly television, very strictly under control.

DIVERSITY

Diversity is an essential requisite of stability. It is no coincidence that in New Guinea, one of the few remaining areas where small tribal societies have not been too severely interfered with, there are seven hundred distinct cultures, each with its own language. To destroy this diversity and set up in its place a monolithic society is to foster social disintegration, with all its attendant problems, which no amount of money or technology can ever begin to solve.

In a country such as Britain, regional differences were once marked. People in different parts of the country had different customs, ate different things, spoke with different accents, and felt correspondingly different. In today's industrial society, an increasingly centralized educational system contributes significantly towards ironing out these very necessary differences and imposing on it a dull and depressing uniformity. This uniformity is not just aesthetically offensive but is also socially disruptive, since it prevents the survival of stable local communities.

DIFFERENTIATION

An individual, in order to be a member of a real society, must also, as we have seen, be a member of a family, age-group, etc.

It is his/her membership of a specific set of such groupings that confers on him/her an identity or status within the society.

In order to fulfil his/her corresponding functions, the precise nature of the information which should be communicated to him/her must of necessity vary with each individual. In this way no two people will receive an identical education.

It might be sufficient to communicate some of this information to him/her relatively late in life. Generally this is not the case for functions requiring great skill. For these there appears to be no substitute for the apprenticeship system. Indeed it is not at the age of nineteen, at a polytechnic, that one can learn to become a master craftsperson or a great artist. It is by diligent and pain-staking work starting at a very early age, during which time valuable information that has possibly accumulated over many generations is passed on, mainly via the family. Needless to say, the apprenticeship system is impossible in a society which regards social continuity as retrograde and social mobility as an end in itself, indeed as the very mark of progress and social just-ice. Such values are indeed very unadaptive, for, by making the apprenticeship system impossible and forcing people into cen-tralized schools, these skills can only be lost and greater uniformity promoted.

Apart from this, and to return to a previous theme, the stabil-ity of the family unit, possibly even more important than that of the community itself, is being effectively destroyed. A child educated far from home in a vast factory-like school and imbued with the values of our technological society, will regard a parent, who still plies a rural craft in the native village, with a mixture of pity, disdain and condescension. To me, this is the ultimate human tragedy, especially as the parent has probably made untold sacrifices to provide the child with that education which he/she believes will ensure social and economic advancement. Yet it is the inevitable concomitant of our highly mobile society for which our educational system is at least partly responsible.

SPECIALIZATION OF SEXES

Specialization in human society is mainly culturally determined. That based on sex however has a genetic basis, though it is fashionable to maintain the opposite. It is one of the illusions of our society with its fixation on uniformity and standardization that men and women are for all purposes, but the sexual act itself, psychologically and behaviourally interchangeable. One must not be fooled, it is maintained, by the fact that they look different: a morphological difference does not imply a behavioural one. It is surprising that apparently serious people are willing to defend so indefensible a proposition. Men and women look different for the very good reason that they are different, and the present-day attempt to iron this out is naive, disruptive and ultimately condemned to failure.

Significantly, in every traditional society known, there is a clear division of labour between men and women. The exact functions fulfilled by the different sexes vary from one society to another. The same theme however runs throughout. Women are responsible for looking after the children for which function they, and not men, are biologically and psychologically adapted. The notion that bringing up the children should be shared by husband and wife takes no account of the fact that only she is capable of feeding the child and also that mother's love is very different from the father's. In *The Art of Loving*, Eric Fromm points out that a mother's love is unconditional.[9] A child can commit the most heinous crimes without it being affected in any way. This is not true of the father's love, which is given on condition that the child behaves itself. This must imply that the child must be psychologically stronger and must probably reach a certain age before it can tolerate the substitution of the father's love for that of its mother.

In nearly all traditional societies other functions undertaken by the women, not directly connected with bringing up their children, tend to be in the home or in its vicinity, while the men wander further afield. Thus in a hunter-gatherer society it is the women who do most of the gathering and the men the hunting.

This does not prevent women from fulfilling all sorts of important social functions and acquiring considerable influence and prestige within their social group. The division of labour between men and women is also important as the relationship between a man and a woman is clearly not symmetrical. They require very different things of each other.

The matriarchal society which people often talk about is rarely found among traditional societies. Many are matrilineal which means that inheritance is via the mother, others are matrilocal which means that a married couple live with the wife's parents. Some are both at once, and in such cases the wife's influence is probably greater than in others, but she still does not run the family.

Matriarchal societies are to be found in the ghettos of the larger American conurbations. Here the men are largely unemployed, incapable of fitting into mainstream society and psychologically prevented from fulfilling their functions as fathers and husbands. Marriage is rare, most unions are temporary, and it is the mother who must take upon herself the responsibility for bringing up the children. This situation is aberrant. Such societies have reached the final stages of disintegration and are hovering on the verge of explosion.

In our industrial society, women tend to be subjected to precisely the same education as men and are encouraged by every means to compete with them. This can only mean the further breakdown of the family unit which depends for its survival on a clear division of labour among its members, which provides the basis for co-operation rather than competition.

Such a policy must also mean condemning our children to a large measure of social deprivation since their mothers, forced to work in offices and factories, sometimes many miles away from their homes, cannot conceivably devote to their children the time required to ensure their proper upbringing. This tragedy can only be reflected in further increases in delinquency, crime, drug addiction, alcoholism and other manifestations of social maladjustment.

It must also mean encouraging the further expansion of our energy and resource-intensive consumer society, since women are thereby ineluctably drawn into the cash-economy and must become increasingly dependent on creches, bottled milk, convenience foods and labour saving domestic appliances.

It must also mean further damaging the health of the population since bottled milk is a poor substitute for human milk, and convenience foods contain potentially harmful chemical additives while most of them have been so devitalized that they but imperfectly satisfy basic human nutritional requirements.

It must also mean creating an increasingly large number of ever more maladjusted women who are being forced by their education and other social pressures to wage an unequal struggle against their natural instincts. It is hardly surprising that more than 50 per cent of teenage girls entering hospitals today are the victims of attempted suicide.

RELEVANCE OF EDUCATION TO LIFE-STYLE

It is an important feature of traditional education that it is totally relevant to the life the young will afterwards lead. The accent is on practical matters but also on the cultural traditions of the social group. The latter's relevance to the problems of everyday life is not immediately apparent to most of us who have been misled into regarding a culture as a random pattern of superstitions and irrational practices. A new approach to anthropology—cultural ecology—is rapidly revealing that a culture is a social control mechanism; that its status *vis-à-vis* a society is similar to that of the personality *vis-à-vis* the individual.

The colonial powers attached very little importance to the cultural patterns of traditional societies and suppressed many of their essential features. Instead, they imposed upon them an alien educational system designed to transform their members into the alienated inhabitants of an anonymous mass society exclusively geared to the dehumanizing goals of mass production and consumption.

In *From Child to Adult* Middleton writes, 'the learning of genealogies of the families and clans, as among the Ashanti and the Baganda, the recognition of social groupings in hierarchical tribal settings and of their reciprocal relationships, the hearing of tribal history in praise songs and legends told at tribal gatherings—these were forms of direct learning which had no place in schools but had set times and places in a traditional situation.'[10] All these activities were not deemed worthy of inclusion in the curriculum of institutionalized western schools.

With the institutionalization of education the tendency is to isolate the educational process from the social one. Rather than be an integral part of it and subjected to the same modifying influence that will enable society as a whole to adapt to changing environmental requirements, it becomes subjected to a different set of modifying influences, of a mainly arbitrary and non-adaptive nature, those imposed by the prevailing values of the day. As Coleman says,

> This setting-apart of our children in schools which take on ever more functions, ever more extra curricular activities—for an ever longer period of training has a singular impact on the child of high school age. He is 'cut off' from the rest of society, forced inward towards his own age group, made to carry out his whole social life with others of his own age. With his fellows, he comes to constitute a small society, one that has most of its important interactions within itself, and maintains only a few threads of connection with the outside adult society. Consequently, our society has within its midst a set of small teenage societies, which focus teenage interests and attitudes on things far removed from adult responsibilities and which may develop standards that lead away from those goals established by the larger society.[11]

It is also true that teachers, both in schools and universities, are increasingly isolated from the real world. The Assistant Masters' Association has complained of just this, pointing out that teachers pass their childhood in the classroom, their adolescence in college and their adulthood back in the classroom. How can they know anything of the outside world?

This tendency can only be accentuated with the further institutionalization of the educative process. Eventually it must destroy the social continuity which is the basis of education in traditional societies. Cultural information, as is already partly the case, will cease to be transmitted from one generation to the next. Instead, each new generation will be called upon to work out its own solutions to problems which are ever more challenging, and whose precise nature they are ever less able to comprehend.

THE DYNAMICS OF EDUCATION

Another of the unfavourable aspects of modern education is that the child's role in the educational process is a largely passive one, whereas when education is part of the social process, as is the case in a traditional society, the child is an active participant in it.

Among the Chaga of Kilimanjaro, as O.F. Rahm writes,

> The child is not a passive object of education. He is a very active agent in it. There is an irrepressible tendency in the child to become an adult, to rise to the status of being allowed to enjoy the privileges of a grown-up . . . The child attempts to force the pace of his 'social promotion'. Thus, at five or six years of age a little boy will surprise his mother one day by telling her that he wants to be circumcised. The mother will hear nothing of it and threatens to beat him if he repeats the request. But the demand will be made with increasing insistence as the child grows up. In former times it was the clamour and restiveness of the adolescents which decided the older section of Chaga society to start the formal education of the initiation camp.[12]

The social process is a dynamic one, and to introduce a child into it is sufficient to ensure its education. Institutions are largely unnecessary. Instead learning is something that takes place like any other behavioural process, so long as the conditions are propitious. As Illich writes, 'By definition children are pupils', and 'learning is a human activity which least needs manipulation by others. Most learning is not the result of instruction. It is rather the result of unhampered participation in a meaningful setting. Most people learn best by being "with it", yet school makes them identify their cognitive growth with elaborate planning and manipulation.'[13] Illich regards the child

> as a typical victim of consumer society and institutionalised education as a specific type of merchandising. The 'school' sells a curriculum—a bundle of goods made according to the same process and having the same structure as other merchandise. The distributor/teacher delivers the finished product to the consumer/pupil, the reactions are carefully studied and charted to provide research data for the preparation of the next model which may be 'upgraded', 'student designed', 'team taught', 'visually aided', or 'issue centred'.

To anyone who has studied the learning process in animals, human or non-human, it is apparent that this type of education is useless. Learning depends on active participation. In addition, the human brain is so designed that it enables information irrelevant to the behaviour pattern to be forgotten. The brain is not a store, but an organization of information, and information is organized in accordance with its relevance. Academic information which does not have obvious relevance to our daily lives is unlikely to be registered more than superficially, nor can it be expected to outlast its usefulness for examination purposes.

THE EXPORT OF EDUCATION

Since we regard our way of life as a model for all other societies, and those who have not yet achieved it as being backward, barbaric and ignorant, we have come to identify education with our particular type of education. It must follow that people who have been subjected to the traditional education of their own non-industrial society are regarded as 'uneducated'. In other words we regard the information imparted in our educational establishments as (a) expressing a set of indubitable truths, (b) expressing the only possible set of indubitable truths, and (c) having universal applicability. A more intolerably presumptuous attitude is hard to imagine, nor for that matter one that is more naive or that reflects a greater ignorance of scientific and social realities.

Needless to say, if education is identified with socialization, then each society must require a different type of education. Thus the programme which will transform a Chaga child into an adult member of its society capable of fulfilling its specific functions within a very distinctive African society cannot conceivably be the same as that which will enable a baby Eskimo to learn its equally specialized, but very different functions, as a member of a family and small community geared to survival in the inhospitable Arctic regions which they inhabit. A Chaga with the education of an Eskimo is, from the point of view of his/her society, uneducated, as he/she would be were he/she to have been exclusively subjected to western education influences even were these to result in a doctorate or a Nobel Prize.

It is evident that the spread of western science-based education throughout the world must lead to the disruption of all other cultural systems. A society subjected to this sort of cultural imperialism finds itself inevitably deprived of the means of renewing itself, since its youth, imbued with a totally alien set of values, will cease, in all but name, to be members of it. Nor will it, in exchange, have become fully transformed into a modern industrial society. It is thus condemned to drift into a cultural no-man's land, from which it can neither retreat nor advance.

It is essential that it be generally understood, before any further irremediable damage be done to the delicate fabric of remaining relatively stable societies, that real education cannot be transferred from one society to another. It is not something that can be imported or exported like cheese or Brussels sprouts. It is only the external trappings of education that can be transferred in this way. Real education must be seen as a process for ensuring the continuity of a cultural pattern and maintaining a social structure and way of life which may have taken many thousands of years to develop as an adaptive response to specific environmental conditions.

CONCLUSION

It is not by further centralizing education or rendering it more capital-intensive that one can combat 'ignorance' or in any way improve the state of the society we live in. Nor is it by forcing more and more of our youth to spend an ever greater part of their lives in the factory-like compounds into which we are at pains to transform our schools and universities. Education is the process of socialization. It is the communication to the young of that information which will enable them to fulfil their functions as members of their families and communities. Ignorance can only be regarded as a deficiency in that process. It is not due to a shortage of educational establishments, to a lack of teachers, nor to a shortage of funds for providing them, but rather to the breakdown of the family and community: the necessary environment of the educational process, without which the latter is but an empty formality. One cannot socialize people when there is no society for them to be socialized into. One must first re-create a society. To do this, one must re-establish those conditions within which the family and community can once more become self-regulating units of behaviour. This basically means de-industrializing society for, with economic growth, the tendency can only be in the opposite direction, as every one of the institutions of an industrial society conspires to bring about further social disintegration. Nor is instruction in modern technologies a substitute for education. It can give rise to a mass society which, for a while, may display a considerable degree of affluence, but not to a structured stable or happy one.

Education is an essential part of the social process. If the latter is deficient, however, then education must be too. Nothing short of a total reorganization of society can provide us with a satisfactory educational system, and this must first of all involve the development of a decentralized society in which each community is allowed to develop its own cultural pattern and its own decentralized educational system for transmitting it from one generation to the next. Only in this way can our youth learn to fulfil its familial and communal functions, only in this way can we develop a society that can renew itself and hence survive.

Notes:

1. Rhodes Boyson, Evidence to the Bullock Committee on Literacy, 1971.
2. A. Williams, President of the National Association of Remedial Education, at Annual Conference, Clacton-on-Sea, 1971.
3. Statistics of Education, School leavers, Vol.2, 1971.
4. Gabriel Carceles Breis, UNESCO Courier, June 1972.
5. Ivan Illich, *Deschooling Society*, London, Calder and Boyars, 1971.
6. Margaret Mead, *Coming of Age in Samoa*, New York, Morrow, 1928.
7. Edward Goldsmith, A Model of Behaviour, *The Ecologist*, Vol.2, No.12, December 1972.
8. J. S. Coleman, *The Adolescent Society*, Illinois, Glencoe Free Press, 1968.

9. Eric Fromm, *The Art of Loving*, London, Unwin, 1957.
10. John Middleton, From Child to Adult, in *Studies in the Anthropology of Education*, American Museum of Natural History, New York, The Natural History Press, 1970.
11. J. S. Coleman, op.cit.
12. O. F. Rahm, *Chaga Childhood*, Oxford, Oxford University Press, 1967.
13. Ivan Illich, op.cit.

CHAPTER THREE

The Ecology of Unemployment

' . . . and here we have our new £2 million electronic digital dole assessor
which does the work of one hundred clerks.'

The Ecology
of Unemployment

C HRONIC UNEMPLOYMENT HAS become a feature of our
socio-economic life, and a growing concern of governments
throughout the world. The only method our society provides for
dealing with this problem however is to stimulate economic
growth. That this does not provide a real solution is clear. Economic growth has been the overriding aim of government policy
for decades and unemployment has continued to rise, particularly in the case of the Third World, to ever more dramatic levels.[1]

This trend is largely the result of population growth, urbanis-
ation and the increased capital-intensity of industry, which
inevitably accompany the spread of western influence. As these
trends occur, so more and more people are brought within the
compass of the cash economy. In the early stages of industrialization it is largely the men who take up employment, the
women tending to stay at home. As development proceeds,
however, and the material requisites for survival in an urban
setting correspondingly increase, so does it become necessary
for women to take up employment as well, increasing further
the demand for jobs.

Let us look a little more closely at the reasons why economic
growth cannot, in the long term, solve the unemployment
problem.

One of its basic requirements is that industry be competitive,
hence machines must constantly be substituted for labour. This is

'economic' because the non-renewable resources and in particular the energy required by these machines are charged, or at least have been up till now, at a ridiculously low price, one that in no way represents their true cost, thereby rationalizing capital-intensity.

By doing this we are continually increasing the need for more capital investment. Also work, by becoming more capital-intensive and hence more 'productive' can be increasingly well remunerated. This is necessary to compensate people for the growing deterioration of their physical and social environment and the increasing monotony of their work. It is also necessary to increase their purchasing capacity, as industrial society can only function if producers are also consumers. In this way, effective demand continues to grow, thereby stimulating further production. The inevitable concomitant of this process is a reduction in the number of people employed for a given amount of capital-investment and for a given level of economic activity.

The capital-intensity of industry, in the meantime, is increasing faster than GNP, which means that every unit of GNP will provide an ever smaller number of jobs. Consider the British chemical industry. It employed 407,000 in 1961 for a turnover of 4,875 million dollars. In 1967 it employed slightly fewer people, 406,000, for a turnover of 7,589 million dollars. Looked at slightly differently, sales were 55 per cent less efficient in providing jobs. If we assume an average reduction in efficiency of 7 per cent per annum then, in this industry, turnover must double every seven years to ensure the same level of employment—which we know to be very unlikely indeed.

In the British textile industry, the number of people employed fell from 719,000 to 584,900 between 1961 and 1968, and this in spite of a marginal increase in investment (from 257 million dollars to 271,300,000 dollars). Employment thus fell by 3 per cent per annum, while investment rose by 0.5 per cent.[2] At that rate, by the year 2106, investment will have doubled while employment will have fallen as low as 18,000.

Since this chapter was written a report by Dr Brian John estimates the cost of constructing a PWR nuclear power station at Trawsfynydd at £1.2 billion. It would employ roughly 300 people to operate it. To quote Dr John 'That works out at £4 million a job, compared with the £20,000 widely accepted as the cost for providing a job in light industry.' The construction of such capital intensive projects can thereby only lead to further unemployment since it reduces the amount of capital available to finance labour intensive projects. As Dr John notes 'since 1971, total CEGB manpower has declined by 20,287, and 108 power stations have been closed.'[3]

It follows that the rate of economic growth required to maintain a reasonable level of employment has correspondingly increased and has now reached a level which we cannot hope to achieve.

An OECD study shows how this must be so in the Third World.[4] It takes as its starting point the average situation in Third World countries: a manufacturing sector employing 20 per cent of the labour force, unemployment and under-employment totalling 25 per cent and an assumed increase in labour productivity of 2.5 per cent per year (the rate achieved in the period 1955-63). On this basis, it is calculated that the growth rate in industrial production necessary to absorb an increase in the labour force of 3 per cent a year (the present rate is 4 per cent per year) would be an annual 18 per cent—as compared with 15 per cent as the best growth rate so far achieved in the Third World. In order to wipe out within a decade the existing unemployment and under-employment, the necessary rate of growth would be 30-35 per cent per year. In fact, industrial output must increase by 3 per cent per year simply to keep pace with improvements in productivity—that is, simply to avoid a reduction in the number of manufacturing jobs available. The report concludes that the 'eradication of general under-employment through the development of industrial employment is a practical impossibility in the medium term.'

Various ILO studies have repeatedly stressed that 'there is not the remotest hope that Western Technology, with its capital-intensive bias, can create the basis for ensuring employment to the over 184 million additional job seekers in India during the next 27 years leading up to 2000 AD.'[5]

Robert McNamara of the World Bank agreed with this conclusion. He pointed out that, whereas an increase in urban employment in the Third World of 45 per cent per year would be 'a tremendous achievement, beyond anything that has been achieved in the past', it would correspond to an increase in the total rural and urban labour force of 1.3 per cent approximately one half of the increase anticipated.[6]

A World Bank report agreed: 'No imaginable rate of increase in industrial and service output' the review concluded 'can absorb the expected supply of workers'.[7] The review stressed that the only solution lay in rural development, i.e. in shifting people back to the villages.

> Progress towards full employment is clearly impossible in the 1970s without a substantial expansion of employment in the rural areas. Integrated rural development as a major priority can be justified almost entirely on the sole basis that it can produce large increases in rural employment. We do not know that it can, but there seems to be no other promising alternative.*

THE AUTOMATION RACE

The tendency towards automation is accentuated by competition between companies and, as the scale of economic activity increases, between countries. Such competition is increasingly taking the form of an automation race which has much in common with the arms race. Thus, even if it had been seen as undesirable to introduce containerization into the Port of London, which meant reducing very considerably the number of dockers employed, there would have been very little choice, since the alternative was to give another port, such as Rotterdam, an advantage which would have enabled it to capture business which would otherwise have gone to London. This, of course, would have had the effect of reducing employment still further. As in the case of the arms race, the automation race can only make matters

---

*Of course the opposite has been done. The introduction of the Green Revolution in many places has ruined small families and forced more people to move to the nearest large conurbation.

worse, since whether we win the race or lose it, unemployment can only continue to rise. It is very much a 'heads you win, tails I lose' situation.

DEPENDENCE ON THE TECHNOSPHERE

In theoretical terms, of course, the basic feature of economic growth is that it involves creating a totally new organization of matter, the technosphere, which is in competition with, and is systematically made to replace, the biosphere of which we are an indissociable part. This substitution has serious implications. For instance, it means that humans are systematically isolated from the natural environment to which they have been adapted by millions of years of evolution, and which has always been capable of satisfying their basic biological and social needs.

Until the Industrial Revolution, people probably never had any difficulty in obtaining food and fresh water nor in finding the material with which to build a shelter, but now they are made to live in vast built-up areas and spend the better part of their day in a factory contributing in some way towards the manufacture of objects unrelated to their personal needs or to those of the family. They are in fact forced into the unenviable situation of having to depend on paid employment in order to purchase the necessities of life, which were previously free for the taking.

The same is true of social needs. People previously lived as members of a family and community and these provided the optimum social environment, that which best satisfied basic social needs. In industrial society however, everything conspires to destroy the family unit as well as the community. Thus, the state largely usurps the functions normally fulfilled by parents. It provides children with free education for instance, and a free health service, correspondingly reducing parental responsibilities. Large companies usurp the mother's functions, since most of the things that a woman would normally have to make for her family, clothes, bread and other staple foods, are now available at the supermarket, while functions that would normally be fulfilled by the children, such as helping in the house, and washing up, have largely been taken over by domestic appliances. The family in industrial society has thus lost many of its traditional functions and has correspondingly disintegrated.

The community fares scarcely better. Mobility is normally such that people are rarely in the same place long enough for strong communal bonds to be established. They usually live in housing estates or residential areas, which are not real communities, since people work elsewhere, often at a considerable distance from their homes. To a large extent, the company a person works for provides him/her with a surrogate communal environment. At work s/he has an identity, which s/he lacks elsewhere. Work also provides an individual with a goal-structure, a sense of accomplishment and a corresponding measure of self-esteem. It is for this reason that unemployment is so intolerable. Even if social security prevents an unemployed person from suffering serious material deprivation, s/he is nevertheless deprived of that essential, albeit surrogate, social environment which a job previously provided. It is almost certainly the psychological effects of unemployment that render so many people in the ghettos of the larger American conurbations incapable of fulfilling their family functions, establishing permanent relationships and taking the responsibility for the upbringing of their children.

EMPLOYMENT IN PRIMITIVE SOCIETIES

It is important to realize that this dependence on paid employment was unknown among our paleolithic forebears, who, let us not forget, are more than 90 per cent of all the people who have ever lived. The reason is simple. Contrary to what is generally supposed, hunter-gatherers were not normally short of food. On the contrary, it appears that they never actually consumed more than about a third of available food supplies very much as do insect populations. Also the amount of work required to satisfy their material requirements was minimal, a few hours a day at the most; the rest of the day being spent in such pursuits as gambling, gossiping and visiting friends. It is not surprising that the very concept of work was unknown to hunter-gatherer societies and that in their various languages one finds no word for it. The gathering of roots and berries and the hunting of wild animals were simply part of a day's *routine*, not to be distinguished from other ways of passing the time, such as gossiping and gambling, and almost certainly equally enjoyable.[8]

Among tribal societies that have given up the hunter-gatherer way of life in favour of pastoralism or subsistence agriculture, very much the same is true. However by advancing that far along the road to 'progress', the amount of work they must do to keep alive increases correspondingly. Animals must be penned and fed, when previously they went free and fed themselves. Fields must be tilled and their produce harvested, when previously food plants grew profusely without human intervention. As Sahlins suggests the amount of leisure decreases as society 'advances'.[9]

Among such societies, trade was as much to reinforce social ties by creating dependencies and obligations, as to satisfy material requirements. People produced essentials for themselves and traded largely what was to them superfluous. Economic activity took place at the level of the family—the basic economic unit, and occasionally families would co-operate to undertake special projects at a communal level. Wage labour did not, and could not exist. 'Nowhere in the uninfluenced primitive society do we find labour associated with the idea of payment.'[10]

Under such conditions, there can be no unemployment in our sense of the term. What can happen, however, is that a population can grow to that point where it can no longer be usefully employed on the land, which would cause surplus people to drift to the cities in search of work. This is in fact what is happening today throughout Africa, but only because of western influence, which has led to the introduction of labour-saving devices into agriculture, and also to the suppression of those cultural controls which previously maintained a check on population growth. It is important to realize that, with the absorption of tribal peoples into the cash economy, they are now not only at the mercy of the vagaries of nature but also of those of the market economy, which our most learned economists are increasingly at a loss to predict, still more to control.

It is also interesting to consider that wage labour only appears in a society where, in the words of Maine, 'contract' has replaced 'status' as the basis of economic obligations.[11] Contract provides a very flimsy basis on which to build a lasting structure, status a very much stronger one.

Status depends on tradition and is associated with mutual obligations. Thus a person could not be deprived of his/her means of livelihood unless s/he committed what was regarded as a sufficient crime against society to justify his/her being ostracized: the direst penalty imposed on tribal members, and this was only possible with the concurrence of the tribe as a whole. To break a contract with an anonymous member of a mass society is far easier. A skilful lawyer is all that is needed.

The manorial system, though much maligned, was also based on status rather than contract. It was very much a system of mutual obligations. The serf could not leave his land, but neither in practice could the lord eject him. It is true that the system was open to abuse since the lord was in a stronger position than the chief, let alone the council of elders, of a tribal society. For this and other reasons, it was undoubtedly less perfect a social system than the traditional tribe, far more perfect, however, than the type of society that replaced it.

It was probably during the thirteenth century that the manorial system broke down in western Europe. This was caused by the development of markets, which in Pirenne's view had been prevented for a long time by the Arab stranglehold on the Mediterranean.[12] This had the most serious consequences. The ecological advantage of the manorial system was that the manor produced for itself only, and did not over-produce and thereby impose too great an impact on the environment, as there was no market where surplus produce could be sold. With the development of markets, production increased dramatically, with consequent deforestation, soil erosion and general environmental degradation.

Society too, was reorganized to satisfy the requirements of production and consumption with the resulting ubanization and social disintegration. Polanyi regards the switch from a subsistence to a market economy as one of the greatest calamities to have befallen western civilization.[13]

RESISTANCE TO WAGE-LABOUR

As we have already noted, one of the principal pretexts for establishing industrial enterprises in areas that have so far escaped their ravages is the provision of employment. It is ironic to note just how strongly people, living in a traditional society, have always resisted being transformed into units of wage-labour. In Assam, nothing would persuade the tribal inhabitants to work on the tea plantations, and labourers had to be imported from already over-populated Bihar. In Ceylon, British planters had to introduce Tamils from southern India to work on the plantations. In the West Indies, the Spanish recruited for this purpose the indigenous Caribs. So little, however, were they suited for this soul-destroying work, that it did not take long for the entire race to become extinct. As a result, slaves were imported, as in North America, from Africa.

Akpapa describes the difficulties encountered by the Enugu coalmining industry in Nigeria in obtaining labour for its mines. Labourers, it appears, had to be press-ganged into working in the mines. Every day, seven hundred of them disappeared never to reappear, unless they had the misfortune to be 'grabbed' again.[14] Eventually the chiefs were employed to force their subjects into working for the mining company and were paid so much for each wage-labourer they provided. Those who refused to obey their chiefs were originally fined. Eventually, however, it became necessary to sentence them to varying periods of hard labour.

This gives some idea of the difficulties involved in persuading people who are leading a perfectly satisfying, self-fulfilling life, to leave their families and communities in order to indulge in monotonous, soul-destroying work in some large enterprise.

CONCLUSION

Indeed, people were perfectly well employed before there was a market economy, before wage labour was ever thought of.

Economic growth is, in fact, not the only means of providing employment but the only means of providing employment of a specific type—capital-intensive employment within the market economy. Though it is probable that such employment is far less agreeable than any other, it does permit, temporarily at least, a higher standard of consumption of manufactured goods. It is only in this way that producers can also be consumers and on a scale sufficient to maintain the industrial machine functioning at the required level. On the other hand, the only reason why people require these jobs and the material goods which they enable them to procure, is that they live in a particular type of society in which the requisites of life must be purchased; in which people are so far removed from their natural environment that they can no longer produce them for themselves, while their essential social environment has been so degraded that they have become dependent on a surrogate one, one that is provided by a place of work, which is divorced from the realities of their family and social life.

The argument that further economic growth is required to provide further employment is only true if we insist on observing the rules of our industrial system, if we remain intent in further extending the technosphere at the cost of the biosphere, and in further increasing the capital-intensity of economic activities. It would not be true if we rejected these goals and opted out of the automation race. The choice is not simply, as we are told, between unemployment and economic growth (with more unemployment at a later date). There is a third course open to us. It is to develop a work-intensive economy in a real and sustainable society. Undoubtedly we would as a result be poorer in terms of largely superfluous material goods and technological devices, but we would be incomparably richer in terms of the social, ecological, aesthetic and spiritual benefits, of which, during the industrial era, we have been so cruelly deprived. There is every reason to believe that it is in terms of the latter rather than the former that our true prosperity should be gauged.

Notes:

1. Jimoh Omo Fadaka, Poverty and Industrial Growth in the Third World, *The Ecologist*, Vol.4, No.2, 1974.
2. OECD, *The Chemical Industry*, *The Textile Industry*, OECD, 1969—70 and 1968—69.
3. Dr Brian John, *The Guardian*, 1 December 1986.
4. OECD, Paris, 1973.
5. F. A. Mehta, *Employment, Basic Needs and Growth Strategy for India*, Geneva, ILO, 1976.
6. Robert McNamara, *Development Review*, November 1972.
7. The World Bank, *Development Review*, 1972.
8. Richard B. Lee and Irven DeVore (Eds), *Man the Hunter*, Chicago, Aldine, 1968.
9. Marshall Sahlins, *Stone Age Economics*, Chicago, Aldine-Atherton Inc., 1972.
10. A. R. Shearer, *The Economics of the New Zealand Maori*, Wellington, New Zealand, Government Printer, 1972.
11. Sir Henry Maine, *Ancient Law*, London, 1861.
12. Henri Pirenne, *Un Histoire Economique de l'Occident Medieval*, Brussels, Desclee de Brouwer, 1951.
13. Karl Polanyi, *Primitive, Archaic and Modern Economics*, New York, Doubleday, 1968.
14. B. Akpapa, Problems of initiating industrial labour in a pre-industrial community, *Cahiers d'Etudes Africaines*, Spring 1973.

CHAPTER FOUR

The Ecology
of Health

'And what did I tell you, Mr Lampsprocket, about leaving the window open?'

The Ecology
of Health

EXPENDITURE ON HEALTH services throughout the industrial world has got out of hand. In many countries it is increasing so rapidly that, at the current rate, it is a matter of decades rather than centuries before it absorbs the whole of the GNP. Clearly, before this point is reached, drastic action to curb health expenditure is required. But how is this to be done? It is generally assumed that the problem is one of organization. Some critics favour the American system of free medical enterprise, others the nationalization of medical services as in the UK, while others favour some intermediary solution such as that adopted in France.

If one looks into the question a little more deeply, it becomes apparent that the problem is not how our health services should be administered but what sort of services should be provided. Those provided today, based on modern medical science, have failed to deliver the goods. If they had been successful, levels of health would be rising and we would expect to see a reduction in the number of people consulting a doctor, in the number of working days lost through illness and in the expenditure on medical services. The opposite is of course the case. As Powles writes 'one of the most striking paradoxes facing the students of modern medical culture lies in the contrast between the enthusiasm associated with current developments and the reality of diminishing returns to health for rapidly increasing efforts.'[1]

It is agreed, of course, that modern medicine has increased longevity, but this has been greatly exaggerated. Dr R Logan, Director of the UK's Medical Research Unit, tells us that 'a man today can expect to live three years more than his counterpart in 1841.'[2] Most of this improvement however occurred before the introduction of scientific medicine. In the period of rapid increase in expenditure on health improvement, life expectancy has more or less levelled off.

Little has done more to increase the prestige and credibility of modern medical science than its apparent success in eliminating infectious diseases, but this success has proved to be short-lived. We are witnessing today a resurgence of infectious diseases throughout the world, in particular of malaria, gonorrhoea, tuberculosis, pneumonia and cholera, while others such as schistosomiasis and dengue fever are spreading to areas where they were hitherto unknown.

The total impotence of modern medical science to reduce the incidence of the so-called 'diseases of civilization'—cancer, ischaemic heart disease, diabetes, diverticulitis, peptic ulcer, appendicitis, varicose veins and tooth caries—is apparent to all. Their incidence, in spite of all efforts made by the medical profession, continues to increase along with per capita GNP.

The only realistic conclusion to be derived from all this is that medical science is on the wrong track, and that a new health policy is urgently required. But what form should it take? It seems clear that, before we can answer this, we have first to rethink basic concepts such as health, disease, medicine and health services.

WHAT ARE HEALTH AND DISEASE?

Health, in terms of our technological world-view, is seen as the absence of 'clinical symptoms', and disease as the presence of such symptoms. But what do we mean by 'clinical symptoms'? What are they symptoms of? Presumably of disease, but this does not get us very far, because many of the diseases we suffer from are classified purely *in terms of their symptoms*. This is true for instance of rheumatism, arthritis and many of the 'diseases of civilization'. It is also true of psychiatric diseases such as psychosis, neurosis and schizophrenia. Often too, the 'symptoms' are but those of the normal workings of the body's defence mechanisms rather than of any really pathological state. As Dr Malleson points out,

> Over millions of years our bodies and those of our ancestors have perfected defence mechanisms against microbial invasions and noxious chemical substances. These mechanisms are very highly developed. For example, mucus which might be dangerous if it were to accumulate in the trachea is expelled by coughing. Toxic substances in the intestines are eliminated by diarrhoea. Microbial invasion of the body is accompanied by a rise in temperature, which is probably intended to increase the rate at which the defence mechanisms are able to act. To suppress this cough, to prevent this diarrhoea, to reduce this temperature is to counteract essential natural processes.[3]

Yet this is precisely what many medical practices aim at achieving. In this way, they mainly serve to eliminate symptoms and, in doing so, tend to exacerbate the diseases they should be serving to cure.

What is more, if the medicines employed are biologically active, they may also produce side-effects and thereby induce diseases where previously there were none. Indeed such 'iatrogenic' diseases, as they are referred to, now make up a substantial proportion of the total disease-load of a modern industrial society.

To treat the symptoms is often futile for another reason. They are often those of a disease that has taken such a hold over an enfeebled patient that, regardless of the medical treatment provided, it must prove fatal. In such conditions, the object of treatment is nothing more than to prolong human life, just for the sake of it, and without any regard for the quality of the life prolonged—an absurd and often immoral enterprise, if we take into account the pain that the patient must suffer as a result of the often drastic treatment needed to keep alive from day to day.

Few people realize what proportion of the national health budget of an industrial country is spent in this way. According to Professor Ross Hume Hall of McMaster University in Canada, 80 per cent of the health budget of that country is devoted to prolonging the lives of patients who, whatever treatment they receive, will die in the next ten months, and in this respect Canada does not appear to be exceptional.[4]

To fight the symptoms of disease is insufficient for yet another reason. The absence of 'clinical symptoms' in a patient cannot necessarily be taken to denote that s/he is in good health. Some 75 per cent of people visiting doctors' surgeries today are said to suffer from no recognizable 'clinical symptoms'. Yet they feel ill and—in some sense of the term—they are ill.

OUR VIEW OF HEALTH AND DISEASE

The trouble is that our health policies are the only ones consistent with the world-view that has developed during the course of our industrial age and which colours our thinking on all the basic problems that confront us today. What is more, they are the only ones that lead to the sale of medical hardware and expertise and that are thus 'economic'. For these reasons, they are the only ones that, at present, we seem capable of entertaining.

The thesis of this chapter is that to understand health and disease we must see them in the light of a very different worldview—one which we can perhaps refer to as the 'ecological world-view'. This involves, first of all, looking at them in a much wider context.

Modern medical science, we must remember, like all the other disciplines in terms of which knowledge has been divided, has been developed on the basis of the experience of the industrial era—a period of about 150 years—which is negligible in terms of humankind's total experience on this planet of several million years. It is in terms of this total experience that we must look at the issue of health and disease.

But this is not sufficient. We think of health and disease as it affects humans, but they are not something unique, they are only one particular form of life among very many. General systems theory, over the last thirty years, has shown that living entities (systems, such as molecules, cells, organisms, ecosystems, etc.) which may outwardly appear to be extremely different are, at a certain level of generality, very similar, and that, at such a level, their behaviour can be shown to be governed by the same basic laws. It also appears that this principle applies to the method by which living systems are controlled. I shall seek to apply a general systems approach to the problem of health.

STABILITY

The tendency today is to see life processes as largely random. Though the notion of randomness is subject to different inter-pretations, it can be construed as designating a state of disorder as opposed to order and goallessness as opposed to directiveness or purposefulness. This is very misleading. Order and directiveness, the latter being really nothing more than four dimensional order, are the most fundamental features of the biosphere. Indeed if the biosphere did not possess these qualities it could not be studied by science whose role must be to establish regularities and patterns whose very presence must imply order and directiveness. Still less could there be a science of cybernetics, the science of control, since to control a process is to keep it on its course or trajectory i.e. in the direction of its goal, the latter concept being taken to be dynamic rather than purely static.

I shall take the goal of life processes to be the achievement of stability. A living system is stable to the extent that it is capable of maintaining its basic structure in the face of possible disturbances. This is another way of saying that it is capable of maintaining its homeostasis (as the term is used by Cannon[5]) in the face of change. This does not mean that a living entity is static, it must change as a means of adapting to environmental changes. But such changes do not occur for the sake of them but as a means of preventing more disruptive changes.

It is only in the light of such theoretical considerations that one can understand what is health. Health in fact can only be seen as the stability of the organism within its social and physical environment. This means that to show that an organism is healthy must mean that it is capable of maintaining its stability in the face of potentially damaging discontinuities. I think that most students of health, who see their subject matter in anything but the very narrow context within which it is presently studied by modern medical science, would agree with a definition of this sort.

It is, for instance, that of Professor Audy who saw health 'as a continuing property potentially measurable by the individual's ability to rally from insults, whether chemical, physical, infectious, psychological or social.'[6] If we define health in this way, then, among other things, our notion of cause-and-effect must be radically modified.

The cause of a disease can no longer be seen to be the immediately antecedent event that triggered it off—the micro-organism for instance that is associated with an infectious disease—but that a constellation of factors has reduced the resistance of the organism to a point at which it falls victim to an insult that would normally induce in it only relatively mild symptoms. If we see cause and effect in this way, then the criterion for determining whether environmental changes can adversely affect health must also be very different from that which is currently applied. It no longer suffices to determine whether such a change actually gives rise to clinical symptoms but *whether it is capable of reducing the overall resistance of living things and hence their stability or health in such a way that they become more vulnerable to other insults.*

The object of a health policy must be equally revised. Rather than waging chemical warfare against the vectors of disease i.e. at eliminating symptoms, it should be aimed instead at creating those conditions in which discontinuities are reduced to a minimum and in which people's ability to deal with such discontinuities is maximized.

On that basis, our notion of what, among other things, constitutes a pollutant must be radically modified. The acceptable level of a pollutant in the air we breathe, the food we eat or the water we drink, is not just that at which clinical symptoms occur, but that which might be considered to have an adverse effect however slight, on our ability to counteract the biological effects of any other insult.

As Professor Samuel Epstein of the Department of Environmental Medicine at the University of Illinois in particular has shown, a very large proportion, perhaps as much as 80-90 per cent of cancers are caused by exposure to chemicals and radiation.[7] Exposure by itself however does not seem sufficient to trigger off a cancer. Thus Professor Bryn Bridges of the MRC Cell Mutation Unit at Sussex University points out that in a healthy body, cells damaged by exposure to chemicals tend to be effectively eliminated,[8] indeed, if we consider the thousands of carcinogenic chemicals to which we are now daily exposed, were this mechanism not operative, there would be very many more cases of cancer than there already are.

If this is so, then the present epidemic of cancer is not only due to the increase in the number of chemicals to which we are exposed, but to the effect of those chemicals and other factors in reducing our ability to eliminate crippled cells.

LEARNING TO LIVE TOGETHER

It can be shown that as systems develop via the evolutionary process, so do they become increasingly stable. Thus a pioneer ecosystem is subject to all sorts of discontinuities. These are slowly ironed out as the ecosystem evolves, i.e. as pioneering species are slowly replaced by more advanced ones and as a 'climax' or adult state is achieved. Climax forests, for example, are subject to few discontinuities. Thus demographic explosions and diebacks which characterize a pioneer ecosystem do not normally occur in a climax forest. Nor do droughts and floods, erosion and desertification.

The same is true of living systems at all levels of organization. As evolution proceeds, they become better adjusted to the particular ecosystem in which they live and hence to the various forms of life that inhabit it.

Thus it is possible to obtain some idea of the time during which an animal has lived in a specific environment simply by determining to what extent it has learned to live with the other forms of life, including the parasites and micro-organisms, that inhabit it. If it has lived in it a long time, then the diseases that could be caused by such parasites have become endemic. They are relatively mild and their function is simply to kill off the old and the weak i.e. to apply quantitative and qualitative controls on host populations.

Consider the case of myxomatosis. It was a well established disease among rabbits in Brazil among whom it is endemic and causes but mild symptoms. It was unknown among European rabbits which are of a different genus. When myxomatosis was introduced into Australia in 1950, the European rabbits introduced there were exposed to a virus of which they had no previous experience. In the first year it killed 99.8 per cent of the rabbit population, in the next year the death rate went down to 90 per cent, seven years later it had fallen to 25 per cent. The rabbit population is clearly learning to live with the virus, and vice versa. The relationship between the rabbit and the virus has thus become progressively more stable.

The same thing has happened to human populations throughout the world, as they have been exposed to parasites of which they have had no previous experience, and with which they have gradually learned to live.

The population of the various islands of Polynesia, for instance, was decimated by the diseases brought there by the European colonists. That of the Maoris of New Zealand fell from approximately 160,000 to 30,000 and at one time it was thought that the Maoris would become extinct. That of Tahiti fell from a similar figure to about 7,000 that of the Marquesas, it is estimated, from 100,000 to no more than about 3,000. However, the Polynesians have adapted to the introduced microorganisms that have become a new component of their environment. They have, in fact, learned to live with them and their population has correspondingly grown. In New Zealand it is now two to three times its former size.

All this makes it clear that as living systems evolve they become increasingly adapted to their environment, and increasingly stable which means that the incidence of disruptive discontinuities is correspondingly reduced. From this must follow the essential principle that the environment which most favours the health of a living system must be that to which it has been adapted by its evolution and with which it has co-evolved.

That this must be so is quite clear in the case of non-human animals. Thus most of us will admit that a tiger has been adapted by its evolution to living in the jungle. It is clearly the jungle that provides its optimum environment. It is the activities it is capable of indulging in in the jungle that best satisfy its physical and psychological requirements. It is the food that it finds there that it most enjoys eating and that best satisfies its biological requirements, and the same must be true of all forms of life, including humans. All must best be adapted to the environment with which they have co-evolved.

The corollary of this principle must also be true. Indeed, as the environment of a living thing is made to diverge from that with which it has co-evolved, and hence, to which it has been adapted, so will it become ever less stable, and hence less capable of dealing with discontinuities, in fact, less healthy.

Stephen Boyden has formulated this principle very clearly. He refers to it as the 'principle of phylogenetic maladjustment'. According to that principle 'if the conditions of life of an animal deviate from those which prevailed in the environment in which the species evolved, the likelihood is that the animal will be less well suited to the new conditions than to those to which it has become genetically adapted through natural selection and consequently some signs of maladjustment may be anticipated.' Obvious though this principle is, and obvious though its importance, it is seldom referred to in the literature, and consequently its significance seems to have been largely overlooked.

> The term 'phylogenetic maladjustment' (the maladjustment is phylogenetic because it represents a characteristic response of the *species* to the changed environmental circumstances) then, specifically refers to disorders which represent the reactions of organisms to conditions of life which differ from those to which the species has become genetically adapted in evolution through the processes of natural selection. This principle relates not only to environmental changes of a physiochemical or material nature, such as changes in the quality of food or air, but also to various non-material environmental influences, such as certain social pressures which may affect behaviour.[9]

THE OPTIMUM ENVIRONMENT

What then are the lifestyle and the environment to which humans have been adapted by evolution and which must thereby most favour the maintenance of human health? The answers, however much we may be loathe to face it, are those of our paleolithic hunter-gatherer ancestors. As Washburn and Lancaster point out 'the common factors that dominated human evolution and produced homo-sapiens were pre-agricultural. Agricultural ways of life have dominated less than one per cent of human history and there is no evidence of major biological changes during that period of time . . . the origin of all common characteristics must be sought in pre-agricultural times.'[10]

It is, in fact, easy to see why the lifestyle and environment of hunter-gatherers should have been so favourable. First of all, such people were constantly on the move which means that they were not for long in contact with their own excrement. This reduced their vulnerability to many parasitic diseases. They lived close to nature and had at their disposal a wide diversity of fresh and uncontaminated foodstuffs. The small groups they lived in were dispersed over a wide area which prevented the spread of diseases from one locality to another. Such small groups, what is more, were not capable of supporting a viable population of the parasites associated with the major infectious diseases that have become current among urbanized populations. A population of 500,000 people, for instance, is required for the measles virus to survive and propagate itself.

Also, since hunter-gatherer groups could survive without disturbing their biotic environment in any way, they did not interfere with established relationships between parasites and their non-human hosts. Bubonic plague, for instance, developed as a disease of rodents, yellow fever and malaria as diseases of monkeys, rabies of bats. Once we destroyed the habitat of the host animals and modified our own so as to create a new niche for the micro-organisms involved, they were quickly transferred to humans. Malaria too is transmitted by the anopheles mosquito which originally preyed on monkeys living on the canopy of tropical forests and to which it was well adapted, causing but mild symptoms in the host. However, once the forests were cut down, the mosquitoes had to find alternative hosts and the most generally available were people.

The creation of vast urban conglomerations has provided a perfect niche for burrowing rodents, including the rats that transmit bubonic plague.[11] It has also put us in close contact with parasites that had previously established a stable relationship with the animals we had domesticated. An example is smallpox, a variant of cowpox, which is a disease of cattle.

Large-scale irrigation projects have also provided an ideal habitat for water-borne diseases. The result is the spread of schistosomiasis and malaria which even the World Health Organisation (WHO) acknowledges to be our doing. 'As he constructs dams, irrigation ditches to alleviate the world's hunger he sets up the ideal conditions for the spread of disease.'[12]

In general, with the development of industry, the environment we live in resembles ever less that to which we have been adapted by our evolution. We are forced to live in massive industrial conurbations which bear little resemblance to the living environment in which we evolved. We live in nuclear families, often truncated ones at that, in a vast atomized society, if indeed we can dignify it with that term—that bears no resemblance to the extended families and other cohesive social groupings within which we have lived over the last few million years.

We eat food that is grown by unnatural processes, making use of a host of chemical substances, hormones, antibiotics, pesticides, including insecticides, herbicides, nematocides, fungicides, rodenticides, etc. of which residues are to be found in practically all commercially available food today. Our food is then processed in vast factories with the result that its molecular structure is often totally different from that of the food we have been adapted to eat during the course of our evolution, and it is further contaminated with thousands of other chemicals, emulsifiers, preservatives, anti-oxidants, etc. designed to impart to it those qualities required to increase shelf-life and otherwise improve its commercial viability.

We drink water contaminated with heavy metals and synthetic organic chemicals, including pesticides, which no commercial sewerage works or water purification plants can effectively remove.[13]

We also breathe air that is polluted with lead from petrol, asbestos particles from brake linings, carbon-monoxide and nitrogen-oxides from car exhausts, sulphur-dioxide from chimney flues, radioactive caesium, strontium and plutonium from nuclear tests, and a host of other radionuclides from the flues of nuclear installations.

It is not surprising that, in such conditions, we should suffer from a whole new range of diseases which, among primitive peoples who lived in their natural habitat, were conspicuous by their absence, nor, in fact, that the incidence of those diseases should vary as it does in direct proportion with per capita GNP i.e. with the extent to which our lifestyle and our environment have diverted from the norm. These diseases are referred to as the 'diseases of civilization', for they are the direct result of a host of changes brought to our lifestyle and environment, which, with increasing development and industrialization, are made to divert ever more radically from those to which we have been adapted by our evolution and which, as Boyden notes, must be the most favourable to the maintenance of our health.

> We can easily think of countless examples of the principle of phylogenetic maladjustment operating in homo sapiens. The traditional 'scourges' of mankind, such as plague and typhus and the great deficiency diseases such as scurvy, beriberi, pellagra and kwashiorker are all straightforward examples of the principle. An examination of reports on the reasons why patients visit their physicians in the most developed countries in Western society today shows clearly that the majority of the disorders of which they complain fit into this category, and are 'diseases of civilisation', in the sense that they would have been rare or non-existent in primeval society (e.g. virus infections of the respiratory and alimentary tracts, peptic ulcers, cardiovascular diseases, obesity, diabetes and probably much psycho-neurosis).[14]

If we have lost sight of this inescapable fact, it is above all because we cannot face its implications. Among other things, *it makes nonsense of the very idea of progress,* which we have identified with development and indeed with industrialization— the last phase of development, which consists in bringing about, as systematically and as rapidly as possible, in the name of improving the welfare of humankind, those very changes that, by their very nature, *must cause our environment to divert as much as possible from that to which we have been adapted by our evolution.*

It is also because of our blind quasi-religious faith in the omnipotence of science and technology, which, we are told, can, among other things, confer on humans the gift of infinite adaptability. But the changes that they permit are only adaptive if this term is used in a very indiscriminatory manner.

True adaptation must refer to changes that counteract discontinuities by creating the conditions that must reduce their incidence and their seriousness rather than merely suppress their symptoms—*changes, in fact, that help increase stability rather than accommodate instability*—changes that create the conditions that favour health rather than suppress the symptoms of ill-health. Most of the changes made possible by medical science are thereby not true adaptations.

Consider the response to the epidemic of tooth decay. Primitive people, on the whole, had wonderful teeth. With economic development, however, the state of their teeth has seriously declined. It is generally agreed that this is the result of eating junk foods—in particular sweets, biscuits, cakes and over-refined white bread. Indeed, it is in Scotland, where such a diet is most firmly established, that people have the worst teeth—40 per cent of young Scots who have attained the age of twenty-five have no teeth at all.[15]

Now the only truly adaptive policy for dealing with this problem, must be to make people change their diet. However, a modern industrial society that sees everything in terms of short-term economics cannot conceivably do this, as it would mean reducing the sales of the food-processing industry. Therefore it has to adopt a different strategy and the obvious one is to engage hosts of dentists to extract rotten teeth and replace them with false ones—a strategy which, like any strategy our society is capable of applying, must serve to further increase economic activity, and hence help create conditions still less favourable to our health and the state of our teeth. This is clearly not a real adaptation but, in Boyden's language, a 'pseudo-adaptation'.

The function of pseudo-adaptation is not to deal with the causes of a disease, but only to mask its symptoms. While, since they are part of the pattern of resource-intensive and polluting economic activities that are making our planet an ever less suitable habitat for complex forms of life, to apply them is, in the long run, *to increase the incidence of the diseases they are supposed to cure.*

The trouble is that just about all our health strategies fall into this category. None of them is truly adaptive, none seeks to create conditions which minimize the incidence of disease, all of them simply seek to apply technological means for masking the symptoms of diseases whose real causes modern medicine cannot address.

TREATING THE WHOLE NOT THE PART

This brings us to another important principle. For a health service to be truly adaptive, for it to treat the causes of disease and hence reduce their incidence and their severity, it would have to treat a population's social and physical environment and its relationship with them and thereby the larger system.

Thus to reduce the incidence of cancer we must above all refrain from exposing our population to all the chemicals that today find their way into the food we eat, the water we drink and the air we breathe. We know that it is exposure to all these chemicals that is a primary cause of cancer. As already mentioned, serious and objective students of this important subject go so far as to attribute 80-90 per cent of cancers to this cause. But to bring about such a change would mean bringing about radical changes to industrial and agricultural practices and indeed to the very priorities of our political and industrial elite, all of which is completely outside the brief of those responsible for our health.

For that reason, the medical establishment in many cases is reluctant to recognise the true causes of cancer. It even refuses to admit that its growing incidence is, in some way, the result of the various changes brought to our lifestyle and our environment by industrial development, and continues to insist that this disease has always been an important cause of death, which is simply not true. It also grossly exaggerates the contribution to the present cancer epidemic of such factors as the consumption of alcohol and of fats and, of course, smoking—factors which are not directly linked with industrialization.

It then correspondingly overrates the ability of modern medicine to treat individual cases of cancer. In reality, there is no evidence whatsoever that either surgery or chemotherapy, the only treatments our society can provide (the only ones that are politically expedient and economically viable) are at all effective. The survival rate of women with breast cancer is the same whether or not they are operated on, while that of people undergoing surgery for lung cancer is less than 1 per cent. In the meantime the incidence of cancer goes on increasing from year to year and whereas it previously mainly affected the middle-aged and the elderly it is now also one of the major causes of death in children.

It is indeed ironic to consider the massive efforts made by our health services to treat so totally ineffectively the ever increasing number of cancer victims when our political and industrial leaders are committed to policies that can only further increase the number of the victims. An effective health service must thereby have an unlimited brief. It must be able to veto government policies in every domain, if such policies can be shown to have an adverse effect on our health. It must, in fact, be able to treat disease at the level of society itself rather than merely at that of the individual.

HOLISTIC TREATMENT

Let us consider some of the obvious advantages of holistic treatment. The first is that the higher the level of organization at which a disease is treated the smaller need be the human intervention—the more the healing process can be assured by the self-regulating mechanisms of nature.

Indeed, the most successful treatments provided by modern medical science are those that seek to create the optimum conditions within which the healing process can occur on its own. When a surgeon stitches up a wound, for instance, or sets a broken bone, that is all that he is doing. The fact is that science is incapable of replicating the incredibly sophisticated biological healing process which is the product of millions of years of evolutionary 'research and development'.

Nor, for the same reasons, can science replicate the social healing process. It cannot transform delinquents, criminals, vandals and drug-addicts, who are largely the products of disintegrated families and communities, into normal well-adjusted adults. The reason is again the same. Socialization is the only means of creating well-adjusted individuals and there is no way in which scientists can artificially replicate it in a disintegrated society, in which socialization can occur but imperfectly. Our only way of dealing with crime, vandalism, drug addicts, etc. is by engaging more policemen, building more prisons, and installing more burglar alarms and other types of anti-crime gadgetry, i.e. once more by suppressing the symptoms of the disease rather than by addressing its causes.

The only way to cure social deviance is to recreate the conditions in which the socialization process can occur, thereby leading to the development of healthy families and communities within which the incidence of social deviance would be reduced to a minimum.

LOGISTICS

What is more, to treat the symptoms of the disease rather than the disease itself, in a routine and systematic way, presents insuperable logistical problems. It means providing expensive technological treatments and capital intensive hospitals for the hordes of people who must inevitably fall sick in the increasingly unhealthy environment of a modern industrial society.

Many general practitioners in the United Kingdom see as many as a hundred patients a day. Studies have shown that the average doctor in the National Health Service writes up to one prescription every six minutes. In such conditions there is no way in which doctors can accurately diagnose the 'causes' of their patients' complaints. All they can do is to dish out biologically active drugs such as antibiotics and cortico-steroids which are likely to have some immediately noticeable effect, hopefully eliminating the patient's symptoms, even though, in the long run, they may prolong the duration of the disease and give rise to all sorts of side-effects.

At the same time, hospitals, in spite of the vast sums of money spent on new ones in the last thirty years, are still incapable of accommodating the increasing number of people who are considered to require hospitalization, and there is today a permanent waiting list of hundreds of thousands of people.

With the inevitable economic decline that faces us today, ever less money is likely to be available for health services and we shall eventually have to face the inescapable fact that it is financially and hence logistically unfeasible to treat disease at an individual level rather than at that of the society and ecosystem whose degradation is its real cause.

Significantly, it is not only ill-health that must be dealt with in this way. None of the basic problems that confront our society today can be solved without bringing about the most radical changes to the society we live in. Take agriculture. There is no sound agricultural policy that could be introduced without changing all the basic features of our industrial society. That we need smaller farms, that they should be geared to poly-culture as opposed to monoculture, that they should adopt sound rotational methods rather than grow the same crop on the same land year after year, all this we know to be true. But to create such farms and enable them to prosper is impossible in our society as it is structured today, and in which overriding political and economic considerations assure that we adopt precisely that form of agriculture that is the least desirable on biological, social and ecological grounds. Indeed, for it to be possible to reintroduce sound agricultural practices, almost everything within our society must change, including family and social structures, life-styles, education, values, fiscal policies, food distribution networks and international trade.

Of course the very suggestion that a Minister of Health, let alone a mere health practitioner, should be able to change the structure of society and its natural environment as the only means of solving the health problems of individual people, would today be regarded as unrealistic if not downright crazy.

But is the idea all that inconceivable? It undoubtedly is if we consider the problem in the context of today's industrial society. However if we view it in the light of human's total experience on this planet, it is seen to be quite realistic. Indeed, in primitive societies, and let us not forget that over 95 per cent of all people have lived in such societies, health was assured in just that manner. Health practitioners (shamans, diviners, etc) maintained the health of their fellow tribesmen by influencing them to act in the way that maintained their human and non-human environment in the state which, among other things, most favoured the maintenance of human health.

Let us briefly see how this was done. A society's behaviour pattern is based on a particular model of its relationship with its environment which is usually referred to as its world-view. The world-view of a tribal society is formulated in a language with which few of us are conversant, that of its mythology, and is concerned with the world of gods or spirits. These are not seen as organized in a random manner, however, but in such a way that the model they constitute reflects people's relationships with their human and non-human environment, on the basis of which, adaptive responses can be mediated.

The spirits can be divided into three broad categories: the first are the ancestral spirits. These have still retained their social identity and are thereby still seen as members of their respective family, lineage group, tribe and society. In this way, their organization reflects, with extraordinary precision, that of their descendants, and serves to sanctify their social structure and hence to preserve it.

Secondly, there are the spirits of nature. All plants, animals and even physical things, such as rocks and streams are regarded as imbued with spirits. In this way, they too are sanctified, which serves to preserve them or at least to reduce the impact on them of people's otherwise destructive activities.

As is now reasonably well known, primitive hunters, before killing an animal, first prayed to its spirit and to the nature god, whose function it was to protect wild animals from depredations, in order to ask for forgiveness for what they were about to do. The totemic system, whereby a particular clan identified itself with a particular animal, also assured that at least in this clan's territory it was regarded as holy and thereby remained unmolested.

Now to sanctify something is the only cultural device that has ever succeeded in preserving it, a fact that is only too easy to verify empirically in the light of the pathetic failure of just about all the efforts of conservationists to preserve our now desancti-fied society and its desanctified environment from our increasingly destructive activities.

A society, however, is not alone in its non-human environ-ment. It is surrounded by other, often hostile, social groups. In addition a society does not always display the ideal degree of order, for not all behaviour within it is under control. In other words, it displays some measure of randomness, and thereby contains some anti-social elements. Such elements, together with neighbouring hostile tribes are represented by the third category, namely the evil spirits and witches.

It is to be noted that the world is not seen as composed of spirits in the way in which scientists see the world as being com-posed of molecules and atoms.

Primitive people do not have a *reductionist view* of the world. The spirits rather than being components of the biosphere are seen, on the contrary, as being organized in such a way as to *re-flect its truly hierarchical nature*. They represent it at every level of organization not just the lowest one as the scientific model does.

A further feature of the primitive world-view is that the inter-relationships that are seen to exist between the different spirits which control society, its enemies and its natural environment, are closely established by tradition and carefully explained in terms of its mythology. What is more, such interrelationships are constantly brought home to people in songs and other ritual activities. Thus, among the Canelos Quichua Indians of Ecuador, as Whitten tells us

> Playing flutes, singing songs and telling myths punctuates discussion of Amasanga (who controls the weather, the thunder and lightning), Nungui (who controls the soil-base for the roots of garden-life and pottery clay) and Sanghui (who controls water). These activities are, among other things, mechanisms for associational, or analogic linking of cosmological and ecosystem knowledge to social rules and breaches, and social dynamics to cosmological premises.[16]

In this respect, the primitive model is also in stark contrast with the scientific model. Rather than being divided up into watertight disciplines between which interrelationships are almost impossible to establish, it is, on the contrary, totally non-disciplinary which, in terms of general systems theory, is required if the model is to permit the mediation of an *integrated behaviour pattern* as opposed to that *mere patchwork of expedients* that is the policy of a modern nation state.

It is to be noted that the primitive model is formulated in a language which all can understand. This too is in stark contrast with the scientific world-view, which we so highly prize and which is formulated in an esoteric tongue which only a handful of specialists can really understand.

This is of particular importance if we consider that stability implies self-regulation. It is cybernetically impossible for a natural system to be governed from the *outside*, for its goal would thereby be *random to it* and, hence, to the biosphere of which it is an integral part, and would simply reflect that of the external agencies that were doing the 'controlling', as it does in our industrial society today.

For a system to be self-regulating means, above all, that the behaviour of each sub-system must be subjected to the control of the system as a whole, and this is only possible if all its members both use and understand the same language, i.e. if the language in terms of which their world-view is formulated be *demotic* rather than *hieratic*.[17]

It is in terms of this world-view that a discontinuity such as a disease is interpreted. Sometimes, it is seen as being caused by the evil spirits that reside in witches or other anti-social elements, or else it is seen as a punishment meted out by the ancestral spirits or the spirits of nature for failure to observe the traditional law and in particular for violating a taboo which, in some cases, is also seen as increasing vulnerability to witches.

A typical example is that of the view of disease entertained by the Luo of East Africa. Among them as Whisson tells us

> Sickness is believed to be caused by spirits falling into different categories, the most current being the spirits of the parents or grandparents (vadzimu), spirit elders or ancestral spirits and the witches (muroi). While

the intervention of the ancestors might be capricious, the diseases ascribed to them or to God were usually felt to be punishments for the sins of the patients or their families. A man who broke a tribal rule might expect to be punished for it by the ancestors or by God in the form of disease. Any man attacked by disease would therefore feel obliged to examine himself and his relationships with the ancestors. A very minor organic disorder—like several days of constipation—might create a considerable overlay of fear or guilt and reduce the patient to helplessness until the rituals were performed and the ancestors propitiated according to the traditions of the society and the directions of the diviner.[18]

In the Old Testament, as we must recall, a natural disaster whether a famine, an earthquake, an epidemic or an invasion by the Philistines was also invariably attributed to failure on the part of the Jews to worship Jahveh in the correct manner, worse still to worship a rival Baal.

The tendency is for people brought up on the modern scientific values to scoff at such a diagnosis. It is 'unscientific' and hence, in terms of our world-view 'irrational', but let us look at it a little more closely. The rules that govern the behaviour of a primitive society that are justified in terms of its mythology and imposed by public opinion, the Counsel of Elders and the ancestral spirits, are not of a purely random nature. They can in fact be shown to be *highly adaptive*.

In the light of the empirical evidence this thesis is unassailable since tribal societies, in particular hunter-gatherer groups, have been able to achieve an unparalllelled degree of stability within their natural environment in which they could have survived and indeed thrived almost indefinitely if their lifestyles had not been interfered with and their environment annihilated by western man. Such stability is maintained by strict adherence to a set of laws that assures above all the preservation of the social and physical environment that most closely resembles that to which the society has been adapted by its social evolution. This being so, failure to observe such laws, the breaking of a taboo, for instance, and indeed sinning in general, can only be construed as a violation of precisely that set of laws that assures a society's success, indeed its survival. To sin is thereby to behave in that way which, in such conditions, must lead to the destabilization of the individual's relationship with his/her society and the society's relationship with its environment, and such destabilization can only be reflected in all sorts of discontinuities of which diseases are but one, this being so, the primitive diagnosis, however quaint the language in which it is formulated, is in fact correct. Indeed, if the behaviour pattern that the gods of a tribal society have sanctified is adaptive, in that it has led to the lowest possible incidence of disease and other discontinuities, then the occurrence of a disease must indeed signify that the society has sinned.

What is more, if the disease, like any other discontinuity, is due to a biological, social or ecological diversion from the optimum, then its cure can only consist in restoring the optimum. This means that if it has been caused by a witch then the activities of the witch must be neutralized, so as to reduce tensions and, at the same time, to reduce those anti-social activities in which a witch may possibly indulge. If the disease is seen as being caused by violating a taboo then the violators must make amends. In particular, they must make the appropriate sacrifices to their ancestral spirits, fulfil their various ritual obligations to kin, cease killing wild animals over and above those that they are ritually entitled to kill and otherwise refrain from doing things which can impair the proper functioning of the society's cultural pattern within its specific environment.

Of course, such individuals are also treated medicinally. For instance herbs and other traditional medicines may be administered as part of a ceremony and these may often prove effective. But to cure the individual is not the prime object of the treatment. It can even be regarded as mere 'spin-off', the real role of the treatment being to restore the biological or psychological stability of the person affected, by restoring the proper functioning of the biological, social and ecological systems whose disruption is the real cause of the problem.

This is the conclusion of Professor Victor Turner with regard to the Ndembu.

It seems that the Ndembu 'doctor' sees his task less as curing an individual patient than as remedying the ills of a corporate group. The sickness of a patient is mainly a sign that 'something is rotten' in the corporate body. The patient will not get better until all the tensions and aggressions in the group's interrelations have been brought to light and exposed to ritual treatment. The doctor's task is to tap the various streams of affect associated with these conflicts and with the social and interpersonal disputes in which they are manifested—and to channel them in a socially positive direction. The raw energies of conflict are thus domesticated in the service of the traditional social order.[19]

It is also the conclusion of Professor Reichel Dolmatoff's study of the way the Tukano Indians of Colombia adapt to their environment. A Tukano shaman, as he shows, does not see a disease as the result of a simple biological insult as would a reductionist scientist, but as a socio-ecological imbalance.

His main concern is about the relationship between society and the super-natural Master of game, fish and wild fruits, on whom depends success in harvesting and who commands many pathogenic agents. To the shaman it is therefore of the essence to diagnose correctly the causes of the illness, to identify the exact quality of the inadequate relationship (be it adultery, overhunting, or any other over-indulgence or waste), and then to redress the balance by communicating with the spirits and by establishing reconciliatory contacts with the game animals. In this way the shaman as a healer of illness does not so much interfere on the individual level, but operates on the level of those supra-individual structures that have been disturbed by the person. To be effective, he has to apply his treatment to the disturbed part of the ecosystem. It might be said then that a Tukano shaman does not have individual patients: his task is to cure a social malfunctioning. The diseased organism of the patient is secondary in importance and will be treated eventually, both empirically and ritually, but what really counts is the re-establishment of the rules that will avoid overhunting, the depletion of certain plant resources and unchecked population increase. The shaman becomes thus a truly powerful force in the control and management of resources.[20]

In this way, primitive people, by correctly diagnosing diseases as the symptoms of social and ecological maladjustment, whether at the level of the individual, the family, the community or the ecosystem, bring about those changes that will put their society back on its correct course; that which will assure a reduction of the incidence of disease to the unavoidable minimum; i.e. to that level at which disease kills but the old and the sick thereby applying quantitative and qualitative controls on a human population so as to help maintain its long-term viability.

It is essential to realize that it is not just diseases but all discontinuities that are interpreted in this way. Droughts and floods and military reversals, as already intimated, are also seen as signs of socio-ecological instability, a thesis that has so far been most forcefully put by Roy Rappaport in his study of the Tsembaga of New Guinea.

> The operation of rituals among the Tsembaga and other Maring helps to maintain an undegraded environment, limits fighting to frequencies which do not endanger the existence of the regional populations, adjusts man-land ratios, facilitates trade, distributes local surpluses of pig throughout the regional population in the form of pork and assures people of high quality protein when they are most in need of it.[21]

As I have already pointed out, this self-regulation requires the concentrated action of the whole society. All its parts must contribute actively. Each individual must be actively involved in the rituals that will assure his or her society's stability.

Our modern industrial society cannot function in this way because it has disintegrated into a mass of unrelated and alienated individuals who do not have the capacity to take a real hand in the running of their affairs. In this way we have become totally dependent on external agents of control. At the same time, our society is so structured that it is impossible to treat a disease at any level higher than that of the individual. Experts in different fields, reared on the specialist knowledge contained within the watertight disciplines into which modern knowledge has been divided, are employed to fulfil carefully defined tasks. Each specialist has a limited brief; he or she cannot venture outside what is considered the legitimate field of activity without venturing on to territory over which some other specialist holds sway.

Not only is this true of the medical profession but even of the Minister of Health. The Minister's territory is well defined. He or she can order the building of more hospitals, subsidize the production of more pharmaceutical preparations, encourage the recruiting of more doctors and nurses. The Minister can also bring about certain changes to the organization of his or her departments, but that is about all. Against the real causes of disease the Minister can do nothing.

What is more it is difficult to see how, within our modern society the present state of affairs can possibly be remedied. Our society is committed to a course—that of further development and industrialization—that can only exacerbate all the basic problems that confront it, including the growing ill-health of its members. In the long-run of course, the problem will be solved, for conditions are becoming ever less propitious to the industrial process, so much so that we are faced in the not too distant future with inevitable socio-economic collapse. It is only then that our health could take a turn for the better, for out of the ruins of our industrial society, we can hope to see emerge smaller, more decentralized societies that might eventually develop the capacity for cultural self-regulation and thereby create conditions that are more favourable to the maintenance of human health as well as that of whatever forms of life may have survived the industrial holocaust.

Notes:

1. John Powles, The Medicine of Industrial Man, *The Ecologist*, Vol.2, No.10, October 1972.
2. R. Logan, Quoted in A. Malleson, *Need your Doctor be so useless?* London, Allen & Unwin, 1973.
3. A. Malleson, *Need your Doctor be so useless?* London, Allen & Unwin, 1973.
4. Ross Hume Hall, personal communication.
5. Walter G. Cannon, *The Wisdom of the Body*, New York, W. W. Norton, 1939.
6. J. Ralph Audy, Measurement and Diagnosis of Health. In: P. Shepherd and D. McKinley (Eds) *Environ/Mental*, Boston, Houghton Mifflin, 1971.
7. Samuel Epstein, *The Politics of Cancer*, San Francisco, Sierra Club, 1979.
8. Bryn Bridges, personal communication.
9. Stephen Boyden, Evolution and Health, *The Ecologist*, Vol.3, No.8, August 1973.
10. Sherwood L. Washburn and C. S. Lancaster, The Evolution of Hunting. In: Lee and Devore, *Man the Hunter*, Chicago, Aldine, 1968.
11. C. C. Hughes and J. M. Hunter, Development and Disease in Africa *The Ecologist*, Vol.2 Nos.9 & 10, Sept/Oct 1972.

12. WHO Chronicle, 1973
13. M. Fielding and R. F. Packham, Organic compounds in drinking water and public health, *The Ecologist Quarterly*, Summer 1978.
14. Stephen Boyden, op.cit.
15. W. W. Yellowlees, *Journal of the Royal College of GPs* 29. 27. 1979.
16. N. E. Whitten, Jr., Ecological Imagery and Cultural Adaptability: *American Anthropologist*, Vol.80, No.4, Dec. 1978.
17. Ken Penney, of Exeter University has suggested the use of these terms in this context.
18. Michael Whisson, Some Aspects of Functional Disorders among the Kenyan Luo. In: Ari Kiev (ed) *Magic, Faith and Healing*. New York, The Free Press, 1967.
19. Victor Turner, A Ndembu Doctor in Practice, In: Ari Kiev (ed) *Magic, Faith and Healing*, New York, The Free Press, 1967.
20. G. Reichel Dolmatoff, Cosmology as Ecological Analysis. A View from the Rain Forest. *The Ecologist*, Vol.7, No.1, Jan/Feb 1977.
21. Roy A. Rappaport, *Ecology, Meaning and Religion*. North Atlantic Books, Richmond, California, 1979.

CHAPTER FIVE

Can Pollution be Controlled?

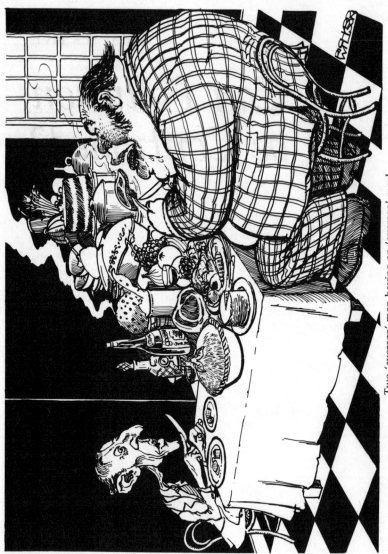

Two 'average' men having an 'average' meal.

Can Pollution be Controlled?

AT THE STOCKHOLM Environmental Conference, in answer to environmentalists' demands for the banning of supersonic aircraft, Lord Zuckerman, ex-chief scientist to the British government, answered that

> If it were an ineluctable conclusion that the use of supersonic civil transport would irrevocably wreck the ozone layer which overlies our atmosphere, can we seriously imagine that we would not find ways of inhibiting the use of such aircraft, as our knowledge of their secondary effects, if any, become more apparent? What are we: ants, lemmings or rational human beings?[1]

Lord Zuckerman appears to be living in a world of his own. Indeed the experience so far has been that very harmful pollutants have often been produced for a very long time before even their harmfulness has been noted, during which time they have entered into so many different manufacturing processes that they have become general environmental contaminants. Even then efforts to introduce any controls at all have been so feverishly opposed by industry and government with the aid of the scientific experts whom they employ that still more time has elapsed before they have met with any success however limited. And limited they always have been. Indeed with respect to the control of pollution we have behaved very much like the legendary lemmings. Let us consider a few examples.

PCBs and DDT

Polychlorinated biphenyls (PCBs) are today regarded as known carcinogens, also ones that appear to be toxic at much lower levels than previously thought. They were first brought into use in 1929. It was nearly forty years before their environmental hazards were recognized. In the intervening years, 30,000 tons had been dispersed in the atmosphere, 60,000 tons into water systems and 300,000 tons had been dumped.[2] If DDT can be transformed into PCBs, which seems to be the case under the action of ultra-violet rays, then the quantities are much bigger still. What is certain is that PCBs are now a general contaminant of our environment. They are present in water, air, soil and sediment, and tend to accumulate in the fatty tissues of animals.[3]

DDT is now also recognized as causing serious biological damage to biological organisms and is a suspected carcinogen. Since it first entered into use the total amount produced is somewhere in the area of two million tons, and, like PCBs, it is now a general environmental contaminant. Though its effects were revealed over thirty years ago with the publication of Rachel Carson's *Silent Spring*, no real action was taken until 1972. Today, even though it is no longer allowed to be sold in the US, exports are unaffected and it continues to be produced at around the rate of 100,000 tons a year.

Vinyl Chloride Monomer

Another general contaminant of life on this planet is vinyl chloride monomer (VCM), which is now a recognized carcinogen. Prior to the discovery of its hazards however, an estimated 100 million pounds a year were being lost to the environment during manufacture and 2 per cent of the US output of five billion pounds was being released 'through deliberate dispersive use', though the most hazardous of these uses, as a propellant in aerosol spray cans, was banned in 1974.[4] The use of this pollutant is now subject to certain controls in many countries, however, its production in total terms has not been affected. In particular it is still used as a plasticizer for wrapping materials and is known to leach in small but significant amounts into the food-stuffs contained.

Asbestos

The carcinogenic effect of asbestos has also been known at least since the early 1930s. The time lag between first reports of asbestos-related disease and control measures to reduce the risk was about thirty years. During this period, to quote Lawrence McGinty,

> The toll of death from cancers and lung disease caused by asbestos will never be counted. Some are buried with conveniently incorrect death certificates, others died from lung cancers indistinguishable from those caused by smoking cigarettes. Although asbestos is not the most potent toxic substance used industrially, its very pervasiveness means that the number of people exposed to it—workers, consumers, and those living in cities—is enormous. It would be quicker to count those who haven't been exposed.[5]

Food Additives

Other carcinogens such as red dye No. 2 or amaranth, another established carcinogen, have entered into an astonishing variety of processed foods. In the US each year about nineteen million dollars-worth of these have been produced and added to between nineteen and twenty-five billion dollars-worth of food. According to the Federal Drug Administration (FDA), amaranth was used in

> ice cream, processed cheese, luncheon meat, frankfurters, fish fillets, shell fish, cornflakes, shredded wheat or wheat cereal, rice flakes or puffed rice, rolls (sweets, cinnamon, Bismark etc.) snack items (spretzels, corn chips, crackers etc.) cookies, pie crust, cake mix, pickles, canned peaches, citrus juice and other canned fruit juices, other canned fruits and fruit cocktails, salad dressings, jelly, pudding mixes, syrup, jam, candy bars, vinegar and cola drinks.[6]

Americans are said to have ingested about five hundred tons of it a year. The FDA attempts to ban it were delayed for fifteen years. It has now been banned in the US but is still in general use in other countries.

Hexachlorbenzene and Hexachlorophane

Hexachlorbenzene (HCB) has been widely used as a fungicide for seed protection. World production is thought to be around four million pounds a year. It was found to be highly toxic as early as 1955 when grain seed in Turkey treated with HCB that was intended for sowing, was used instead for bread production. Five thousand people were affected by eating the contaminated loaves and between 250 and 500 died. WHO has shown that children under the age of two taking HCB via their mother's milk suffered a 90 per cent mortality rate. In the US, HCB is a trace contaminant of human milk and levels are also to be found in other food stuffs including butter. So far efforts to ban this substance have failed.[7]

Hexachlorophane is also a highly toxic chemical. In the summer of 1972, thirty-nine infants in a French hospital died from being rubbed with a baby powder containing 6 per cent hexachlorophane. It was banned by the FDA in January 1972, but only after being used for thirty years in a host of non-prescription products including 400 categories of deodorants, soaps, shampoos, toothpastes, cleansers, and cosmetics, involving thousands of brand names and hundreds of millions of dollars in retail sales.[8]

Chloroform

Chloroform was shown to cause liver cancer in small animals over thirty years ago, but it is still used in cough mixtures, mouthwashes and toothpastes. It is also used as a preservative. In fact it is now in such general use that when it was suggested to a UK expert that this substance might be banned he answered 'it would be like trying to get rid of alcohol—there is a little bit everywhere.'[9]

It is important to realize that I have named but a handful of the four million chemicals which, according to OECD, we have introduced into our environment; 563,000 of these are thought to be in common use and a hundred are produced in excess of 50,000 metric tons a year.[10] It is also important to realize that very few of these chemicals have been properly tested—a question I shall examine in greater detail further on in this chapter.

It should thereby be fairly obvious that we live in a highly contaminated environment—and when we consider that possibly 1,000 new chemicals are introduced every year—some say 3,000—and that the quantities of the existing ones generated by our industrial activities continue to increase with the growing world economy, our environment is clearly becoming more highly contaminated every year—which explains the current and growing epidemic of pollution-induced diseases, in particular cancer.

What is more, it is equally obvious that our efforts to control pollution are very ineffective—that we are, infact, behaving much more like lemmings than like Lord Zuckerman's 'rational human beings'. Why should this be so?

THE EXCUSES

When the pollution caused by a particular activity or set of activities is pointed out to the polluters or to the authorities and it is suggested that something be done about it the answer is nearly always the same. We are told that this is not possible until further scientific research is undertaken in order to obtain 'the hard scientific evidence' required to determine the exact effects of the pollutants on living things.

Consider, in this respect, the efforts made by Chemie Grunentaile (the firm responsible for making and distributing thalidomide) to avoid the banning of this product. Among other things a famous expert, Professor Eric Blechschmidt, Director of the Institute of Anatomy of Gottingen University, was prevailed upon to state in a Court of Law that 'so long as there is no complete certainty about how the thalidomide might execute its effect on any embryo, theories about the drug's detrimental quality are premature and represent no more than pure speculation. Any binding thesis about a causal link between thalidomide and deformities does not yet exist.'[11]

Consider too the efforts made to control the pollution of the Mediterranean.[12] The United Nations Environmental Programme sponsored a series of international meetings to try to reach agreement on actions required to prevent this sea from becoming a lifeless waste but so far these have been in vain. The excuse for inaction is, as usual, ignorance as to the exact nature of the pollutants that are causing all the damage. 'We do not have enough evidence yet', said Dr Keckes, the scientist in charge of 'the scientific assessments' of the state of this highly polluted sea. 'The rate of implementation of the Treaty', he states 'will depend on how fast we can define the standards scientifically.'[13]

Consider yet another example. The National Cancer Institute (NCI) has demonstrated the carcinogenicity of trichlorethylene which is used as an industrial solvent particularly for degreasing machine parts, in foodstuffs for decaffeinating coffee and has been used by solvent abusers with fatal results. In the UK the Health and Safety Executive (HSE) however refuses to reduce existing safety levels which are at present 100ppm. One reason is that the NCI tests were on animals and not on humans. This is a weak excuse since substances which are carcinogenic to one form of life tend to be carcinogenic to others—the informational medium contained in the genes and in the nucleus of a cell affected by a carcinogen being expressed in precisely the same medium, DNA. Tighter controls, they insist, will have to await the result of surveys of workers exposed to this chemical. What is more a spokesperson has admitted that the government Employment Medical Advisory Service *had no immediate plans for such a survey*.[14] Indeed whenever we are told that no scientific evidence has been found to incriminate a particular chemical, the chances are that no one has bothered to look for the evidence.[15]

The question we must ask is how much can governments justify allowing several million chemicals to which organisms and ecosystems have never been exposed during the course of their evolution, to be released into our environment without such tests being carried out? Greater irresponsibility is hard to imagine. But would carrying out such tests serve any real purpose? Would they yield serious information that could really be of use in determining an effective strategy for controlling offending chemicals? As I shall now attempt to show, the answer is undoubtedly no.

ACCEPTABLE LEVELS

The theoretical basis of pollution control is provided by the principle that dangerous chemicals are only dangerous when used at sufficiently high levels, and that there must be a level at which concentrations of the chemical are biologically harmless. If this is so then it suffices to ensure that these levels are never exceeded for no biological damage to occur.

The more we know about the biological effect of chemicals, however, the more it becomes apparent that this is simply not true. Serious efforts have been made to establish safe levels of different pollutants, but these studies have always proved to be in vain.

This is the case for instance with radiation. It is now generally accepted that any increase in radiation levels during the course of our evolution must be reflected in some biological damage. The same seems to be the case with asbestos. The US National Institute of Occupational Safety and Health (NIOSH) has stated quite explicitly that 'excessive cancer risks have been demonstrated at all fibre concentrations studied to date. Evaluation of all human data available provides no evidence for a threshold or for a "safe" level of asbestos exposure.'[16]

In general this is true for all carcinogens, a principle that is accepted by the Health Education and Welfare (HEW) in the US. As its former secretary, Arthur Flemming, has said, 'Scientifically there is no way to determine a safe level for a substance known to produce cancer in animals.' It is this principle that provides the rationale for the famous Delaney Clause which makes it illegal in the US to add any chemical to foodstuffs that can be shown to be carcinogenic even in very small amounts. Since the Delaney Clause was passed no significant progress has been made in our ability to determine a safe level of a cancer-causing chemical. On the contrary, according to Anita Johnson 'recent evidence suggests that estimating a safe dose is more difficult than was previously thought.'[17]

For instance the FDA conducted a massive 'megamouse' study at a cost of more than five million dollars to determine if very low doses of a known carcinogen over a period were in any way safer than high doses, as is generally maintained by the chemical industry. Twenty-four thousand animals were used in this study and the results showed that low doses were not safe at all. Liver cancers were produced at the lowest dose as well as at the highest.

If this is so, then the acceptable levels fixed for potentially dangerous chemicals released into our environment have no scientific basis whatsoever. In fact, it is very easy to show that they are simply *the minimum levels that can be achieved without compromising economic priorities.*

Thus the WHO standard of 0.02 to 0.05 ppm mercury in food 'is simply the practical residue limit, the concentration of mercury expected in the diet from natural background and environmental contamination.'[18] In Sweden, the maximum permissible level was maintained at 1.00 ppm for a very long time. The reason is that if the limit of 0.5 ppm had been adopted in the late 1960s as was then proposed, it would have become necessary to close down more than 45 per cent of Sweden's inland fisheries.[19]

The acceptable level of lead in drinking water is above that which tends to be present today. When WHO recently raised it from fifty microgrammes to a hundred microgrammes per litre, this was not the result of the sudden discovery that man was less sensitive to lead poisoning than was previously thought, but because *few water authorities could provide water to this standard.*

Indeed the Cox report, drawn up with the aid of experts from twenty-four countries, established that toxic effects are noted above fifty microgrammes per litre—while Dr Kehoe who, according to Professor Bryce-Smith 'has probably carried out more studies on lead than almost any other authority living', argued very strongly that the level should be set as low as twenty microgrammes per litre.[20]

The level fixed by the FDA for aflatoxins, highly carcinogenic substances produced by certain moulds left on improperly dried crops such as peanuts and cereals, is 20 ppb. Yet we know that aflatoxins can cause cancer in animals when fed at levels of 45 ppb to fish, 1 ppb to rats; while at 15 ppb aflatoxin, 100 per cent of rats get cancer. As Anita Johnson points out, these very high tolerances are an indication of the commerical pressures felt daily at the FDA.

In the same way, the level set by the FDA for PCBs is 5 ppm in fish and poultry in spite of the fact that PCBs are so toxic that they have been shown to suppress reproduction in animals at a dose of 2.5 ppm. To quote Anita Johnson again, 'The FDA appears to have chosen 5 ppm simply because it would permit the vast majority of PCB-contaminated products to be marketed as usual.'

Efforts to reduce permissible levels by various government agencies are consistently being thwarted by commercial pressures. The Department of Labor, for instance, was expected to announce a reduction in the acceptable level for occupational exposure to lead in the air by fifty microgrammes per cubic feet of air. This would have reduced the existing level by four times. The US Regulatory Analysis and Review Group opposed this on the grounds that the cost to industry of maintaining these levels would be in excess of a billion dollars. New levels for exposure to benzene proposed by OSHA were also successfully quashed in the Federal Appeal Court in New Orleans on the grounds that the agency had failed to demonstrate 'a reasonable relationship' between anticipated benefits and costs.[21]

INTERNATIONAL VARIATIONS

Needless to say, acceptable levels vary considerably from one country to another, largely because of the different pressures exerted by government departments and commercial interests.

Thus, the acceptable level for Vinyl Chloride Monomers (VCMs) was for a long time 550 ppm in the US whereas it was only 100 ppm in West Germany and only 10 ppm in the USSR. This seems to have been largely because of the pressure applied by Goodrich, the principal manufacturer in the US, which strongly opposed the proposed new standard. (It was then reduced to 1 ppm over an eight-hour period with permitted excursions of up to 5 ppm over and above which respirators must be worn.)

This refusal to take any real action to ban carcinogens reflects standard government policy. Indeed the government boasts of its 'flexible and pragmatic approach' whereby polluters are simply asked to keep pollution levels down 'by the best practical means' which, in practice means, in that way that interferes as little as possible with the far greater priorities of maintaining employment and economic growth.

Thus the International Committee on Radiological Protection (ICRP) recommends that all exposure to radioactivity 'be kept as low as reasonably achievable, economic and social factors being taken into account; and . . . the dose equivalent to individuals should not exceed the limits recommended in the appropriate circumstances by the Commission.' As Professor Athersley pointed out at the Windscale Inquiry, what this implies 'depends on how "reasonably achievable" is interpreted.'[22]

At Windscale (now Sellafield), where levels of caesium and alpha emitters such as plutonium and americium to the Irish Sea are very much higher than they are from any other similar reprocessing plant, and where no efforts whatsoever are made to contain emissions of krypton 85 and tritium to the atmosphere, the term 'reasonably achievable' is clearly interpreted in such a way as to make it unnecessary for British Nuclear Fuels to incur any additional costs that might reduce this state-owned company's economic viability. This attitude is reflected in all our government's pollution control policies. Thus the UK representatives at EEC discussions on the control of pollutants released into European rivers refused to accept the controls proposed on the grounds that our rivers were faster flowing and could therefore absorb more pollutants than could those on the continent. This is an absurd argument since the levels of pollutants entering our grossly polluted seas from our rivers, are in no way affected by the river's rate of flow.[23]

INDIVIDUAL VARIATION

The most obvious problem involved in fixing an acceptable level for any pollutant is that susceptibility to different chemical substances varies from individual to individual, even more so from one age group to another. As Professor Schubert points out, the reason is often genetic and is related to the presence or absence of the enzyme needed to break down a chemical. Children are particularly susceptible to most pollutants, foetuses even more so. Thus whereas adults appear to take up 10 per cent of the lead that enters their body, the rest being eliminated, the figure for infants may be as high as 50 per cent.[24]

The susceptibility of foetuses to damage by X-rays is well known. During the first three months of pregnancy when the organs are developing they are more susceptible to mutations and a dose of 500 millirads appears sufficient to double the number of cancers likely to develop within the first ten years of life.

The doubling dose however appears to increase to about 2000 millirads if exposure takes place during the last six months of gestation when the foetal organs are growing. In general, cells undergoing rapid division as in the foetus and new born baby are most vulnerable to carcinogens.

Individual variation is also related to domicile. People are more or less vulnerable according to their proximity to a source of pollution, and of course it will vary even more in accordance with the work they are involved in, which may bring them into contact with pollutants of varying degrees of toxicity. Individual lifestyles are also very relevant, obesity and alcoholism may increase susceptibility to pollutants while eating and drinking habits affect resistance to chemicals that may find their way into different types of food and drink. Diet can affect susceptibility to pollutants from whatever source because malnutrition weakens all the body's defences. Unfortunately all such variations in susceptibility to different pollutants are simply ignored.

Thus in the US, the FDA established an 'interim guideline' of 0.5 ppm of mercury in food. This is based on the assumption than an average serving of fish is 150 to 200 gms. It also assumes that the *average* American does not eat fish more than twice a week, and, as a result, would not consume more than 0.03mg. of methyl mercury per day which is the maximum safe limit.[25]

This may be all right for the average American, but *everyone is not average*. What happens to those people who happen particularly to like fish, or who live in an area near the sea where a lot of fish happens to be eaten, or who are professional fishermen, or married to fishermen? Is it right that they should be condemned to suffer the terrible consequence of mercury poisoning?

In Britain there is the same callous disregard for the non-average consumer. Thus, towards the end of 1970, studies began to show that fish at the top of the food chain such as tuna and sword fish, wherever they seemed to come from, were heavily contaminated with mercury. Nine hundred million cans of tuna were taken off the market in the US because they contained more than 1 ppm of mercury. In Britain, tuna on the market was found to contain similar levels of mercury. For a while, the public was advised not to eat it, but then, as Anthony Tucker points out, the government 'contrived to rationalize a course of inaction by resorting to averages. By counting British heads and the number of cans sold a year, and by completely ignoring those who like tuna and eat a lot, the scientific advisory committee was able to arrive at reassuring conclusions.'[26]

Mr Prior who was then Minister of Agriculture made this announcement at the end of the year,

> The experts consider that it is the total intake of methyl mercury that is important . . . they consider that it is unnecessary at this time to withdraw from sale or to advise consumers not to eat any of the canned tuna now on the market. The tests have shown that all this fish is within the safety limit set in some other countries including Sweden. In short, there is no reason why the housewife should not buy it.

CRITICAL GROUPS

This iniquitous philosophy is clearly reflected in the UK in the concept of 'critical groups' i.e. of people who are particularly vulnerable to a specific pollutant. It is on the basis of the exposure of such groups, for instance, that the permissible levels for the discharge of radioactive wastes are calculated.

One such critical group, living in South Wales, is composed of local eaters of traditional laver bread, made from seaweed, which unfortunately tends to concentrate radio-isotopes of various sorts and in particular ruthenium. Concentrations in the seaweed in the vicinity of the Sellafield nuclear installation are already above the permissible levels (in a recent test they averaged out at 370 picocuries per gramme of seaweed). This, however, is not regarded as a cause for concern for the unbelievable reason that the Cumbrian seaweed comprised only a small proportion of that used in laver bread manufacture. Too bad of course for anyone who should wish to use more Cumbrian seaweed than the average for making laver bread. But the situation is now even worse, for a few years ago the women who used to collect Cumbrian seaweed stopped work. As a result the laver bread eaters of South Wales are no longer considered to be a 'critical group', and the pollution of seaweed with radioactive waste is no longer of any concern.

If this principle were generally applied to determine permissible levels of all the pollutants released into our environment, we would be safe so long as we were average consumers of each contaminated foodstuff and, of course, so long as *we did not change our habits in any way from year to year*.

At the same time, if ever political, social, economic, ecological or climatic changes of any kind forced us to adopt new living, working, eating, drinking or breathing habits, we would then be exposed to dangerous levels of all sorts of pollutants that we would have previously not encountered in conforming to the authorities' model of the average person.

In other words, this present system involves constantly narrowing down our options for the future and hence reducing our capacity to adapt to changing conditions: a truly suicidal prospect. If our government were really concerned about the effect of pollutants, it is not for the critical groups of today that it would cater *but for all the possible critical groups of an unpredictable tomorrow*. This means that we should not be catering for the average person but for the maximum person.

CATERING FOR THE MAXIMUM PERSON

There are two other very good reasons why in any case this must be so. Firstly, even if it is possible to be average with regard to exposure to a single pollutant, it is obviously not feasible to be average *with regard to the four million or so pollutants that we have introduced into our environment.*

In reality every single person is *likely to be non-average, in some respects at least,* and *so to be subjected to more than average levels of one, and more likely, many pollutants. The fact is that the average person does not exist. S/he is but a figment of the statistician's imagination.*

The second reason why we must cater for the maximum person is that humans are not the only form of life on this planet. We cannot regard as expendable, and thereby justify systematically poisoning, all the non-human forms of life that happen to inhabit the areas that humans do not exploit to satisfy their immediate requirements. We can only do so at the cost of causing serious ecological disruption with all sorts of unpredictable consequences. Unfortunately, of course, to reduce pollution levels to the point required to protect the maximum person and living things in general from pollution damage, could only be done by compromising other priorities. The cost to industry would be too great—economic growth could no longer be maintained.

MEASUREMENTS

A further problem is that it is difficult to fix levels below those that can actually be measured by currently available techniques, yet many pollutants may well be biologically active below this level. Thus, Dr Sturgess of the Essex River Authority points to the damage done by a hormone weed killer in concentrations as low as one in a thousand million. 'Present methods', he writes, 'do not even enable one to detect the presence of pollutants in this dilution, let alone trace them.'[27]

The measurement of ozone concentrations in the stratosphere which are required to confirm or invalidate the thesis that it is being depleted by fluorocarbons and other chemicals emitted by our industrial activities are only possible in a rudimentary way. 'Available measuring techniques are too insensitive to detect small long term variations.'[28]

Asbestos, even at levels of 0.1 fibres per cubic centimetre of air (the acceptable level proposed by the US National Institute for Occupational Safety (OSHA)), is very difficult to detect and may still cause cancer, since the average worker doing an eight-hour shift would be breathing about 800,000 fibres into his or her lungs every day. The difficulty in measuring asbestos pollution in the environment is illustrated by the experience of the now defunct Greater London Council's Environmental Sciences Group. Air samples taken by this group were sent to eight different commercial laboratories for analysis. Each one gave a different set of results.[29] In the same way during the course of a study carried out by the French consumer magazine *Que Choisir*? a sample of human faeces to which a specific pathogen had been added was submitted to thirty-two commercial laboratories for analysis. Only one succeeded in detecting its presence.

SUB-LETHAL EFFECTS

For these and similar reasons it is only feasible to measure high levels of specific pollutants in our environment and low levels are simply assumed to have little effect on the health of human and non-human animals. This, of course, as already intimated, is a pure act of faith, based on no evidence of any kind. On the contrary the more we learn about the biological effects of different pollutants, the more it becomes clear that exposure to low levels over a long period, can be as damaging, if not more so, as exposure to high levels over a brief period. This is certainly the conclusion of Dr Waldichuck of the Pacific Environment Institute, as he showed in a paper he presented at a Royal Society Meeting on the effect of sub-lethal levels of pollution on marine organisms.[30]

Waldichuck points out that actions to control pollutants in rivers or the sea only tend to be taken after fish kills, i.e. to deal with acutely toxic conditions. It may be just as important to guard against low levels of pollution which, among other things, can adversely affect fish reproduction and thereby insidiously assure the decline and indeed the disappearance of fish populations. These are very difficult to control, indeed, fish populations or even species, can disappear in many cases without anyone noticing it, all the more so in that natural fluctuations in fish abundance and changes due to fishing may obscure a decline in population caused by a pollutant.

There is already ample evidence of fish populations declining and even disappearing from polluted fresh water environments. Thus species of fine fish have declined very dramatically in the North American Great Lakes. The Atlantic salmon has disappeared from many polluted streams in Europe and eastern North America. The Pacific salmon has been affected in the western United States, while populations of other fish have declined and, in many cases, disappeared from acidified lakes in Scandinavia.

The sub-lethal effects of low-levels of different pollutants are best understood once we have adopted an ecological view of health.

Today, individuals tend to be regarded as healthy to the extent that they do not display clinical symptoms of disease, but this is totally unrealistic. An organism must be regarded as healthy to the extent that it is viable, more precisely, to the extent that it is capable of dealing adaptively with environmental challenges. In this sense, health is synonymous with homeostasis or stability. It can be shown, as I pointed out in the previous chapter, that health can be impaired by all sorts of minor insults, in that these can reduce an organism's ability to deal with environmental challenges. This appears to be the view of Waldichuck. He shows many of the ways in which a marine organism's behaviour pattern and hence its ability to survive is impaired by low doses of different pollutants. Thus, very small quantities of cadmium affect calcium metabolism with adverse effects on the equilibrating mechanism of fish.[31] This probably reduces the ability of fish to avoid predators and also their capacity to seek and capture their prey.

Very low levels of various pollutants, in particular of organophosphorus pesticides, inhibit enzyme activity, and can also impair hormone function. Different concentrations of copper and iron have been found to affect plasma cortisone and other hormones in Sockeye salmon.[32] Low levels of oil pollution have been shown to have serious effects on fish eggs, giving rise to chromosomal errors and gene-level mutations, and, when these occur during the gastrula stage they are almost invariably lethal.

Low levels of pollution also give rise to behavioural abnormalities. Normal schooling behaviour, for instance, can be disrupted by a pollutant. The learning ability of fish can be affected by very low concentrations of chlorinated hydrocarbons.[33] This may impair their ability to return to their home stream, especially if they are subjected to these concentrations in the juvenile stage during the imprinting period. Low levels of pollution can also affect the chemoreceptors of fish which also impair their ability to find their home stream, to locate food and avoid predators. A fish's equilibrium can also be affected by low levels of mercury.[34,35,36]

Sockeye smolts infected by parasites succumb to lower concentrations of metals than uninfected fish and, in general, fish exposed to low level pollution either become infected by a disease more readily than unexposed fish, or may break out with a disease that previously existed only in a latent form. What is important is that in general an organism whose health has been impaired by low levels of different pollutants need not display any clinical symptoms until such time as the cumulative damage makes it vulnerable to a particular insult, which in normal conditions it could simply have taken in its stride. This means that by the time clinical symptoms do appear, the organism is already so badly damaged that it may no longer be viable.

It is easy to see why this must be so once one understands the physiology of basic biological functions. To quote Anthony Tucker

Functions such as seeing or the co-ordination of movements or, indeed, any activity, are not the outcome of the activity of single neural cells. All are the outcome of processes which involve thousands and probably millions of interconnected cells. Such systems have what electronic engineers are prone to call, a high level of redundancy. That means that many pathways are not strictly necessary for the function to be carried out efficiently, but simply duplicate or triplicate other pathways in case some kind of blockage or breakdown occurs.[37]

This consideration needs weighing very carefully when considering the effects of organic mercury, lead or the many organochlorines and other pesticides which, like DDT, can incapacitate neural systems, for it means that by the time clinical symptoms of nervous disorder appear, such as lack of co-ordination or loss of vision, *then enormous damage has already been done to the structure of the brain*. It also means that quite extensive damage can be done without any clinical symptoms ever becoming detectable. This, in relation to heavy metals, is a terrifying prospect.

Dr S.G. Rainsford of the medical branch of the British Factory Inspectorate, comes to a similar conclusion when reporting on a six-year survey of lead workers carried out by his organization. 'The most disturbing factor,' he writes, 'is that the worker can be severely poisoned by lead without either having symptoms or showing clinical signs of plumbism. Probably the commonest early symptoms are abdominal discomfort, dyspepsia, loss of appetite and general aches and pains, the latter frequently being described as rheumatism, fibrositis, etc.'[38] Another very significant study also cited by Anthony Tucker was carried out in Edinburgh in 1965. It revealed that symptoms of lead poisoning tended, in the course of medical practice, to be treated as a normal occurrence without there being any attempt to seek their cause. Needless to say, the health of those affected had been impaired and the chances of their succumbing to a serious disease, which they could otherwise have easily survived, had correspondingly increased.

The same principle is well illustrated by the well publicized incident in which fifty thousand to a hundred thousand sea birds, mainly guillemots, died from no apparent cause in the Irish Sea in the autumn of 1969. Largely because of the popular outcry, the incident was carefully investigated by a team of scientists headed by Dr Martin Holdgate, head of the government's Central Unit on Environmental Pollution. What is particularly interesting is that no single factor could be incriminated. During the summer of that year there were heavy storms which may have made it difficult for the birds to feed themselves. Indeed all the dead birds seem to have suffered, in different degrees, from starvation. Their bodies also were found to contain unusually high levels of PCBs. These tend to accumulate in the body fat, just as does DDT, to which these substances are closely related. It seems probable therefore, that, being short of food, their surplus fat was mobilized which led to the transfer of these poisons into the blood stream. This could quite conceivably have resulted in their death. But further examination showed that their bodies also contained relatively high levels of all sorts of toxic elements such as cadmium, selenium, mercury, lead, etc. which could also have contributed to their demise.

Anthony Tucker points to the implications of this disaster:

> The ragged wreckage of dead birds speaks for itself. Like similar wildlife catastrophies in North America and elsewhere, this disaster contains a grave message whose importance cannot be overestimated. It is that the thresholds of environmental calamities are obscure; that levels of contamination are already past the point at which they can amplify many times—perhaps hundreds of times—the fatal effects of purely natural stresses; that it is impossible to predict where disasters will strike and often impossible to define causes after they have happened; and that there is little to be gained from niggling arguments about particular effects of individual components of contamination. The burdens of toxic metals as a whole, and of organochlorines, have already degraded the entire context of the lifesystem.[39]

In the meantime, judging from reports in the *Marine Pollution Bulletin*, the contamination of sea birds and hence of the marine environment itself around these islands continues to increase. A 1973 study on a representative series of marine and estuarine birds revealed that they were all grossly contaminated with mercury. Levels in their liver ranged from 0.7 ppm to 122 ppm in the case of a redbreasted merganser.[40] We can only understand the full horror of these revelations if we realize how many other equally toxic pollutants are likely to be lodged in the livers of these birds—pollutants which our scientists have not yet been engaged to look for and measure.

What is happening to our sea birds is almost certainly also happening to us. A population living in a hideously contaminated environment such as ours, must be a sickly one, which in fact it is. It is not surprising that there are hundreds of thousands of people waiting for beds in our hospitals and that the cost of our National Health Service should now be nearly £19,000 million (1987) and increasing, much faster than the GNP. Indeed, if current trends are allowed to continue, the whole of our GNP in a few decades from now, will have to be spent on trying to reduce the toll of human disease. In the meantime the tendency in diagnosing a disease is still to look unsuccessfully for single cause-and-effect relationships. It is assumed that if a person is ill there is a single reason for it. The possibility that the illness might be due to the combined effect of hundreds and thousands if not millions of different 'causes' all of which have contributed to reducing his or her resistance to disease, is not even considered.

SYNERGISTIC EFFECTS

It is not just the effects of so many pollutants that we must take into account, but the possible synergistic effects between specific pollutants which, as the SCEP Report states, are 'more often present than not.'[41]

Thus Professor Irving Selikoff has shown that asbestos insulation workers who smoke cigarettes have an eight times greater risk of contracting lung cancer than other smokers simply by virtue of their exposure to asbestos, but a ninety-two times greater risk than non-smokers because of the very powerful synergic effect between exposure to asbestos fibres and tobacco smoke.[42]

The addition of mercury to sea water appears to inhibit the growth of twenty-one different strains of bacteria involved in the degradation of oil in sea water.[43]

The presence of a small amount of DDT, equivalent to that found in humans, greatly increases the liver damage produced by small amounts of carbon tetrachloride.[44] The toxic effects of this solvent are also increased a hundred-fold by the addition of the common drug phenobarbital.

The addition of oil to water substantially increases the damage done by DDT, PCBs and other such poisons that are not very soluble in water, but that are very soluble—perhaps as much as 10,000 times more so—in oil.[45]

The addition of the commonly used dispersant BP.1100 × Finasol OSR 2 to oil during a spill increases its toxicity to herring larvae by fifty to one hundred times, quite apart from seriously increasing the period during which it is acutely toxic.[46]

The use of chlorine as a disinfectant in our drinking water may have a whole range of synergistic effects with chemicals that are often present in it. Thus when associated with benzene it can, in the presence of ultra-violet light, give rise to hexachlorocyclohexane (HCH)—a particularly toxic insecticide.[47]

Epstein points out that modern toxicology does not take into account additive or synergic effects.[48] Nor are they taken into account, as Schubert points out, in setting the acceptable levels for exposure to mutagens.[49] For instance, the International Commission for Radiological Protection defines an acceptable risk as one which would involve the doubling of the spontaneous rate of occurrence of genetic damage. But if one takes into account the possible synergistic effects of each mutagen, then such a criteria seems absurd.

As Professor Bryn Bridges notes

What is a suitable recommendation for one mutagen (i.e. radiation) will not suffice when each of a number of mutagens is considered. It has been estimated that about a 1,000 to 1,500 new chemicals are introduced into the environment each year, of which no more than a minute fraction is tested for mutagenic activity. If a thousand mutagens were each allowed at population doses which doubled the spontaneous rate, then the overall rate might go up a thousandfold, quite apart from any synergistic interaction which might occur.[50]

Of course, if minute doses of different pollutants are damaging, then one can only reduce the damage by limiting the number of pollutants released into our environment. Such a policy, of course, would not be consistent with the achievement of our present economic priorities.

TUMOUR PROMOTERS

It also appears that cancer can be induced by a single exposure to a very low level of a known carcinogen, one that would not normally cause cancer, when this happens to be combined with prolonged exposure to equally low levels of a substance that is not in itself carcinogenic. Such a substance is referred to as a tumour promoter.[51] This principle may explain the induction of skin cancers, and has also been shown to be relevant to explaining other cancers, including those of the lung, colon, bladder and liver. It tends to confirm the notion that carcinogenesis is a multi-step process. A single insult may not always be sufficient, others of a different sort being required before cancer develops. Croton oil is said to be a promoter. It is a complex mixture of chemicals including esters of the plant alcohol phorbol. The ester referred to as TPA (12-0-tetradecanoylphorbol-13-acetate) is an especially effective promoter. It appears that the combination of an initiating carcinogen plus TPA is at least ten times as effective in inducing tumours as the carcinogen alone. Other suspected tumour promoters are phenobarbital and the artificial sweeteners saccharin and sodium cyclamate. Bile acid is meant to be a promoter which is supposed to explain, in part at least, the relationship between cancer and a high fat diet.

How the promoters act is not entirely clear. One effect may well be to interfere with the normal development of cells. This is shown to be the case with phorbol ester promoters by researchers at the Wistar Institute of Anatomy and Biology at the University of Pennsylvania. These promoters inhibit the differentiation of a variety of cells in culture. Once the cells become mature, they lose their capacity to divide, but if their differentiation is inhibited, they remain in an immature state and continue to divide, perhaps in an uncontrolled way as do malignant cells.

If this is so, then the implications are dramatic. It means that in trying to determine whether a particular chemical is carcinogenic in the real world, we must not only look for synergic effects with other carcinogens, but also with a host of other non-carcinogenic chemicals which could conceivably act as tumour promoters.

DECAY PRODUCTS

To fix an acceptable level for any chemical would also require taking into account the nature and toxicity of the substances that, under different conditions, tend to be associated with it, which would include, as Epstein points out, its 'chemical and metabolic derivatives, its pyrolytic and degradation products and its contaminants and reactor products.'[52]

Thus the thinning of the shells of bird eggs which has resulted in large-scale breeding failures and the near extinction of many species of birds at the top of the food-chain appears to be caused not by DDT as suspected but by its decay product DDD. It is also possible that under certain conditions, it degrades into PCBs which would make nonsense of calculations of acceptable levels of this substance in industrial effluents.

Cyclamates are not dangerous in themselves but because they decay into carcinogenic cyclohexylamine. NTA, which detergent manufacturers introduced to replace phosphate-based detergents, also breaks down into products with toxic properties.

Mercury in its inorganic form is largely insoluble in living tissue. Its half-life in most mammals, as Anthony Tucker points out, is only six days, which means that it is rapidly removed by the body's natural detoxification mechanisms. However under the action of bacteria in the water and in the soil it breaks down into an organic compound whose half-life in the body is about seventy days.[53] This means that the total body burden of someone ingesting two milligrammes of inorganic mercury would be no more than twenty milligrammes however long the exposure. In the case of organic mercury the figure would be 200 milligrammes within a year, though death would intervene long before.

The same appears to be true of lead, which scientists have discovered to be very readily transformed into an organic form which is very much more toxic than the inorganic variety. It is probably also true of plutonium. Dr Bowen has reported several times that in the marine organisms examined in his laboratory, he found that concentrations in living tissue were a hundred to several thousand times higher than in the surrounding water.[54]

Particularly important for those eating fish from the Irish Sea is that the plutonium isotope plutonium 241 can decay into americium 241. The latter is an alpha emitter—the former is not—and releases of it are thereby not included in the permissible levels of alpha emitters released into the sea from the Sellafield retreatment plant in Cumbria. In 1969, the permissible releases were increased from 1800 curies to 6,000 curies per year and since then, according to Peter Bunyard, forty times more plutonium 241 has been discharged from Sellafield than all the alpha emitting isotopes of plutonium put together. It may well be that

> British Nuclear Fuels who operate Sellafield have succeeded in conning the public into being allowed to discharge more than seven times the quantities authorized which are already scandalously high, as evidenced by the fact that plutonium, one of the most carcinogenic substances known, is now a general contaminant (though at the moment in small quantities) of fish life in the North Irish Sea.[55]

Once again we are faced with a consideration that unfortunately we cannot afford to take into account without compromising economic goals. The cost of examining the innumerable decay products of all the chemicals we have introduced into our environment would be incalculable, the consequences of banning the chemicals whose decay products proved to be harmful, economically inconceivable.

TOXIC IMPURITIES

Toxic impurities also tend to be present in many chemicals causing them to be very much more toxic than they would otherwise be. This is known to be the case with the organophosphate, diazinon, used in homes and gardens for cockroach control. It contains an impurity called sulphotep which is thirty to a hundred times more toxic than diazinon and is much more stable. Its build-up in the environment is favoured by the fact that organophosphate pesticides tend to be applied at more frequent intervals than the more persistent chlorinated hydrocarbons that they have in some cases replaced, which must lead to a gradual building up of sulphoteps.

DELAYED ACTION

A further complication is that the effect of pollutants on biological organisms may take a very long time to show up. Thus seven years after atomic bombs had been dropped on Hiroshima and Nagasaki, a high incidence of leukaemia began to be observed among the survivors. Because after a few years this began to fall off, it was assumed that the worst was over. Fifteen years or so later, however, an unusually high rate of cancer started appearing among the survivors. This tended to confirm the now well-established fact that tumours may appear a very long time after exposure to a carcinogen.

It has also now been found that some cancers only appear a generation later. A rare form of vaginal cancer, for instance, has been observed among women whose mothers were administered the hormone diethystilboestrol (DES) during pregnancy. Carcinogens are also known to cross the placenta and affect the foetus causing cancer in later life.

The delayed effects of pollutants in increasing the vulnerability of ecosystems to population explosions, in reducing the resistance of populations to diseases and also in seriously changing the composition of the atmosphere and the stratosphere, may take a very long time to detect. As the OECD Environmental Directorate reports, for instance, 'because of the long time-lag between release of fluorocarbons to the atmosphere and their migration to, and eventual removal from the stratosphere, their full impact may not be apparent for a decade or more.'[56] This may be an understatement since it may even take as long as a century for halocarbons to reach the stratosphere.[57] The trouble, as the Environmental Directorate points out, is that by that time it may be too late to avoid serious consequences for humans and their environment.

The delayed effect of carcinogens also renders useless epidemiological studies of carcinogenesis triggered off by substances that have not been in use for at least twenty to twenty-five years. Unfortunately, a large proportion of potential carcinogens, in particular synthetic organic chemicals, fall into this category, as do cyclamates. The Committee of the National Cancer Institute looking into the safety of this substance concluded that no epidemiological data were available since bladder cancers had a latency period of twenty years or more and 'cyclamates had not been on the market long enough for cancers to show up.'[58] Yet we are still constantly being assured of the safety of chemicals on the basis of such tests.

As Wurster pointed out, the studies conducted by Hayes and his colleagues and those conducted in the Shell Laboratories which claimed to establish the innocuousness of DDT could not have done so for a variety of reasons, one of which was that the periods of exposure were too short to detect carcinogenecity.[59] Yet the National Cancer Institute's report on the effects of fluoridation of American water supplies states that 'no significant excess mortality from cancer could be detected up to fifteen years after fluoridation in areas where ninety-five per cent of the population had been abruptly and continuously exposed.' This is supposed to justify the dubious thesis that fluoridation does not increase the cancer rate.[60]

THE IDENTIFICATION OF CHEMICALS

The problem of establishing levels for different environmental pollutants is further complicated by the fact that we have identified no more than a fraction of them.

As Professor Rene Dubos pointed out, there are probably hundreds of unidentified pollutants in car exhaust alone. He estimates that we have identified less than 30 per cent of those contained in the air we breathe in modern cities, but 'recent experiments,' he writes, 'have shown that newborn animals exposed to these undefined contaminants may show disastrous consequences when they become adults.'[61]

The plume of vapour from the Seveso Plant, as Anthony Tucker notes, probably contained 'a cocktail of great complexity whose constituents were not only biologically potent at concentrations close to the limits of detection . . . and ''so subtly interwoven chemically'' as, for all practical purposes, to defy identification.'[62]

An EPA study of America's drinking water reveals that 'there may be a myriad of organic chemicals, not yet isolated and identified, such as the pesticides that could be present in these water supplies, which are carcinogenic, teratogenic or mutagenic.'[63]

In the UK, a study by Fielding and Packham comes to the conclusion that it is, to all practical purposes, impossible to identify the organic pollutants in our drinking water. Even though a litre of urban drinking water may not contain more than twenty milligrams of organic constituents, this small amount of material 'is a very complex mixture containing hundreds of different compounds some of natural and some of synthetic origin. Its analysis is difficult and even the most advanced and elitic technology cannot yet identify more than ten to twenty per cent of the organic material present.'[64]

The inter-agency Task Force on Inadvertant Modifications of the Stratosphere (IMOS) has pointed out the difficulty in identifying stratospheric pollutants. 'The additive effect from several substances', it warns 'might become significant in the future, even if the effect from any individual substances is relatively small.' It also warns that there may well be materials yet to be invented or discovered that are serious candidates for concern.[65]

DIFFERENT LEVELS AT DIFFERENT TIMES

The problem of establishing acceptable levels and monitoring them satisfactorily is further complicated by the fact that levels observable in an organism or in an ecosystem are constantly changing. Thus it has been found that American oysters accumulate twice as much cadmium from the surrounding water during the months of July and August as during the winter and the spring. The reason appears to be that the higher water temperature during the summer increases the oysters' metabolic rate. This causes them to pass more of the surrounding water over their gills in a given time and thereby to be exposed to more cadmium.[66]

The DDT content of barracuda in US waters which tends to be high most of the year, falls by about 75 per cent during the spawning season. It coincides with the loss of fat in which the DDT is stored and is somehow returned to the environment, but just how, nobody seems to know.[67]

Unusual climatic conditions can also lead to big variations in pollution levels. The 1976 drought in Britain, by drastically reducing the flow of water in British rivers, radically reduced their capacity to dilute pollutants. This could not but affect fish life and also reduce the quality of available drinking water.

Because of the 1976 drought too, ozone levels in central London regularly rose above twenty parts per 100 million. Worse still, in early July, for eight consecutive hours on five consecutive days, ozone levels averaged more than ten parts per 100 million (the industrial safety level) with peaks of twenty-five parts per 100 million, which was well above previous peaks of sixteen pphm at Harwell.[68]

The ozone concentration of the stratosphere is also changing very regularly. 'There are large natural variations of ozone on a scale of days to many years, and of a complexity which cannot be readily incorporated into current predictive models.' The natural variations are in fact so large that it has been 'estimated that a five to ten per cent decrease in ozone, persisting and measured for several years, would be needed before a change could be attributed to man's activities with any statistical reliability.'[69]

Levels are also constantly changing simply because of people's polluting habits. If a chemical company cleans out vats containing some noxious chemical on a particular day, levels of the pollutant which may have been very low the day before, will clearly substantially increase, at least temporarily.

Dr J. Sontheimer, a chemist working on the pollution of the Rhine notes how levels of different pollutants vary from day to day. 'There is no way of foreseeing what will be floating in the river tomorrow', he writes and as a result 'a cleaning process that works well one day, works badly the next.'[70]

Under these conditions the value of individual measurements is negligible. To be significant, they would have to be carried out over a long period—which, among other things, would present insuperable logistical and financial problems.

ACCIDENTS

Establishing acceptable levels for different pollutants is in any case fairly fruitless in view of the increasing vulnerability of our society to large-scale technological accidents that are already leading to the exposure of whole populations to very high levels of dangerous pollutants.

Government and industry invariably assure us that the odds against a serious accident occurring in a particular industrial installation are massive—one in a million, for instance, against this occurring at a nuclear power station. However the odds against two jumbo jets colliding on a runway appeared to be equally negligible until it actually occurred at Las Palmas. So were the odds against any of the large-scale pollution disasters of the last few years, the massive leakage of radioactive waste at Hansford, the Dioxin disaster at Seveso, the arsenic trioxide disaster at Manfredonia, the PBB one in Michigan as well as the Chernobyl disaster (which, of course, has occurred since this chapter was written). The experts told us they could, in effect, never happen—but they all did and the damage they have done has been on an intolerable scale.

MURPHY'S LAW

We must realize that it is impossible to design, build and operate a technological device that cannot go wrong. This is a principle known to engineers as 'Murphy's Law' which states that 'if something can go wrong, it will.' One cannot ever design a type-writer, a bicycle or even a motor car, that is not subject to break-downs. This may not matter too much, for any accidents that can occur to such devices are on a relatively small scale and will affect but a small number of people. This is not so, however, in the case of accidents occurring at nuclear and modern chemical plants.

Apart from technological breakdowns one must also consider the human element. People simply cannot be counted upon to deal with routine matters, day in day out, with the care and attention normally displayed in emergency situations only, which means that accidents caused partly at least, by human error, are, in the long run, inevitable.

Consider the case of the famous accident at the Browns Ferry Nuclear Power Plant in Alabama on 2 March 1975. Mr Gregory Minor, the manager of advanced control and instrumentation, stated himself that the safety systems in operation 'went far beyond the normal levels of reliability.' What happened was that fire destroyed the safety equipment. This sort of mishap was indeed extremely unlikely but could not be ruled out. 'You can't expect these things to run flawlessly for forty years with so many people involved' he said.[71]

With the rapid breakdown of our society and an ever more reduced sense of responsibility, the problem is likely to get worse rather than better. It is interesting that a chief engineer at Sellafield should be the one to make this point:

> I do not think the country can operate with an acceptable standard of safety an extremely dangerous plant like Windscale under current standards of respect for law, national and personal morals and discipline in social and industrial affairs. To maintain safety in such a plant calls for standards of personal dedication, sense of responsibility and discipline which do not generally exist in the permissive society. This has been demonstrated by the fact that the Windscale workforce was prepared to hazard public safety in pursuit of a minor financial objective.* Might not a later generation occupy the plant and threaten sabotage if their demands are not met?[72]

*It went on strike

The nuclear installation at Sellafield has already been the scene of many accidents, one of which in 1957 was very serious, leading to the escape of considerable quantities of radioactive gases into the environment. Dr Wakstein made a study of the lesser ones from the limited amount of material made available by British Nuclear Fuels Ltd (BNFL). He could only find reference to twenty-eight accidents. However, prompted by the then Minister of Energy, Tony Benn, BNFL later produced a list of 177, most of them additional and as Wakstein points out 'not hitherto disclosed to the public'.[73] If the overlap in the second list is eliminated, it appears that there was a total of 194 accidents and 'incidents' between 1950 and mid 1977 and more have occurred since. Eleven involve fires or explosions, seven have reference to criticality and about forty-five involve releases of plutonium, the average rate is seven or eight a year and the number is increasing. BNFL refuse to regard them as accidents. They are referred to instead as 'incidents' and on each occasion the public is assured that no harm can possibly come to it. This is a totally dishonest assurance, since the pollution released cannot be permanently isolated from the biosphere and must somehow and sometime find its way into biological systems thereby causing biological damage including mutations and cancers. What must be considered is the sheer number of these incidents. Also imagine what it would be like in this country if we went ahead with Mrs Thatcher's programme and sought to become dependent on nuclear power stations for 50 to 80 per cent of our energy requirements, which would mean covering this tiny island with a network of several hundred nuclear power stations together with their allied installations, etc. We must also imagine what it would be like if we then went ahead with our breeder-reactor programme.

Sir John Hill in his time the most fervent advocate of nuclear power in Britain admitted that 'if something went wrong with a fast breeder reactor it could explode. No plant of any description' he said 'can be made to deliver over a million horsepower without the chances of an explosion if something goes wrong.'[74] Dr Farmer, Safety Advisor to the UK Atomic Energy Authority, has admitted that if this occurred there could be as many as a million casualties.

Dr John Edsall, the Nobel Laureate, also describes what the dangers from accidents might be if the US went ahead with its breeder-reactor programme: 'The hazards of the present reactors will be multiplied many fold in the breeder; an explosion in a fast breeder could make thousands of square miles uninhabitable for many years, and could endanger the lives and health of millions of people.'[75]

THE EPIDEMIOLOGICAL APPROACH

It may be argued that we can make up for the difficulty in establishing acceptable levels of a particular chemical by concentrating more on epidemiological studies. These can obviously help, but too much cannot be expected of them in view of the fact that people living in modern industrial conurbations are exposed to a wide range of different pollutants at levels which, in a given area at least, may not be too dissimilar, and that, in such conditions, the identification of a dangerous chemical is only likely under exceptional conditions. Thus, as Epstein points out, the carcinogenic effect of asbestos was determined largely because of the very rare form of lung cancer (mesothelioma) associated with it. The carcinogenic effect of diethystilboestrol (DES) was also only discovered because of the very rare form of cancer of the vagina (adenocarcinoma) it induced in the daughters of women exposed to it during pregnancy—and even then it would probably not have been discovered if a lift had not broken down in Boston, enabling the paths of a gynaecologist and a pathologist to cross for a sufficiently long time for them to exchange relevant experiences.[76] The teratogenicity of thalidomide was recognised only by the bizarre deformities it produced. 'In all likelihood', as Epstein writes 'thalidomide would still be in use as a safe drug had it produced relatively common anomalies, such as cleft palate or strial septal defects.' Epstein further points out that no known major human teratogens such as X-rays, German measles, mercury or thalidomide 'have been identified by prospective epidemiological approaches, even in industrialised countries with good medical facilities.'[77]

A further problem is that adverse reactions to drugs are rarely reported. Thus, though there were recurring complaints about the side effects of the drug practolol which ICI has now taken off the market for causing sclerosing peritonitis, these tended to be ignored by prescribers even though they were often serious (damage to sight, hearing, or the gastro-intestinal tract).[78]

FROM ANIMAL TO HUMAN

Even were we to overcome all these problems we would still be faced with a further one. For obvious reasons, it is very difficult to establish permissible levels on the basis of experiments with humans. Animals of some sort must be used. Unfortunately, however, tests carried out with laboratory animals only provide a vague indication of how they will affect humans. Epstein points out, for instance, that

> meclozine, and antihistamine, used to treat morning sickness, is teratogenic to the rat, but apparently not so to humans. The opposite is the case with thalidomide, to which humans appear to be sixty times more sensitive than mice, a hundred times more than rats, two hundred times more than dogs and seven hundred times more than hamsters.[79]

On the other hand, because of the basic similarity of all forms of life at a molecular level, it seems reasonable to consider that chemicals which are carcinogenic to one form of life tend to be carcinogenic to others as well.

THE NUMBER OF POLLUTANTS

The problem is further aggravated by the literally incalculable number of chemical substances we have introduced into our environment.

To begin with, there are those that have been introduced on purpose—and are used in commercial products of different sorts. Their number is increasing very rapidly every year. According to Blodgett, 400 active chemicals were used in this way in 1965, formulated in over 60,000 registered products of which some 35,000 were for agricultural application. By 1973, however, the number of registered products was 33,000 incorporating some 900 different chemicals, although twenty substances amounted to 25 per cent of the US production.[80]

As Blodgett points out, very few of these have yet been adequately tested. In the meantime an estimated 500 to 2,000 new chemicals enter large-scale commercial use each year. Of these, the NCI subjects only 150 to long-term rodent feeding tests, each of which in 1976 was said to cost 100,000 dollars.[81] But we must also take into account the far more numerous by-products generated during the production of these chemicals. Together, according to the United Nations Environment

Programme (UNEP), they amount to several millions and the number of further substances that combinations of these could yield is so great as to defy the imagination.[82]

Dr Saffiotti of the National Cancer Institute estimates that two million at least are known. Of these, however, terrifying as it may seem, only 3,000 have been adequately tested for carcinogenic properties, while 1,000 have shown some signs of being carcinogenic.[83]

But what do we mean by adequate testing? The NCI tests are carried out on an average of 800 animals. But can such tests really provide the information we require? Undoubtedly not. If we take into account the immense number of potentially harmful chemicals to which industrial people are exposed, and their additive and possible synergistic effects, we must test for minute biological effects which, needless to say, renders the problem even more intractable. As Epstein writes,

> Assume that man is as sensitive to a particular carcinogen or teratogen as the rat or mouse. Assume further that this particular agent will produce cancer or a birth defect in one out of 10,000 humans exposed; then the chances of detecting this in a group of fifty rats or mice, tested at ambient human exposure levels, are very low. Indeed, samples of 10,000 rats or mice would be required to yield one cancer or teratogenic event, over and above any spontaneous occurrences; for statistical significance perhaps 30,000 rodents would be needed.[84]

MEGAMOUSE EXPERIMENTS

Saffiotti considers that to test potential carcinogens at very low levels similar to those at which human populations may be exposed through residues in food, for instance, and in order to detect a low incidence of tumours, about *100,000 mice would be required per experiment*. Each experiment would cost about fifteen million dollars (1979) and to carry out a significant number of them would 'block the nation's resources for long term bioassays for years to come and actually prevent the use of such resources for the detection of potent carcinogenic hazards from yet untested environmental chemicals.[85] Even then, the results, for a number of reasons, would be highly contestable. To begin with such an approach assumes that there is a threshold dose at which a carcinogen is no longer effective and, as we have seen, and also as Saffiotti points out, 'there is presently no significant basis for assuming that such a threshold would appear.'

Secondly, these studies would, in any case, have to be confirmed by other tests carried out in different conditions such as variations in diet, variation in the vehicles used, in the age of the animals, in their sex, etc. Each of these tests would then imply further megamouse experiments. What is more, they would clearly have to be tested in combination with countless other chemicals, with which they may have additive or synergic effects. They would also have to be tested over that period during which delayed symptoms might be expected to occur, while to test for mutations would mean carrying out such tests on animals for many generations.

THE ECOLOGICAL APPROACH

The fact is that the *problem cannot be solved in terms of what passes today as 'scientific method'*. This is now admitted by a growing number of scientists who have seriously considered all the factors involved. Professor Alvin Weinberg is among them. He considers that a new 'trans-scientific' methodology is required for this purpose. On this subject it is worth quoting him in full.

(The question) What is the effect on human health of very low levels of physical insult can be stated in scientific terms; it can, so to speak, be asked of science, yet it *cannot be answered by science*. I have . . . proposed the name trans-scientific for such questions . . .

Let me use as an example of a trans-scientific question the problem of low-level radiation dose . . . One may well ask, assuming the dose-response curve to be linear down to zero dose, how large an experiment would be required to *demonstrate* empirically that 170 millirems . . . would increase the mutation rate by the 0.5 predicted by the linear dose-response theory. The answer is that around 8×10 mice would be required to demonstrate a 0.5 per cent level at the ninety-five per cent confidence level. So large an experiment is beyond practical comprehension. The original question as stated is therefore, in my terminology, trans-scientific. Where low level effects are concerned, there will always be a trans-scientific residue.[86]

It follows that it is not by making millions of deceptively precise measurements that we can understand how pollution is affecting our environment, such an enterprise being, among other things, logistically impossible. It is the effect of pollution *taken as a whole* on living systems *taken as a whole* that we must consider. This is the conclusion of Caroll Wilson and his colleagues' 1969 Study of Critical Environmental Problems (SCEP) *Man's Impact on the Global Environment*, which is still by far the best study on global pollution problems. Its authors considered that our 'total pollution burden may be impossible to determine except by direct observation of its overall effects on ecosystems.' This is also Schubert's conclusion.

It has become apparent that an overall approach is necessary if society is to control and minimize genetic and toxicological risks to the population. It is unproductive and self-defeating to repeatedly deal with an individual chemical on an emergency basis simply because it happens to make the newspaper headlines. Repetition of such piecemeal consideration eventually distracts the public and government from the general problem of how to deal with the myriad of chemicals to which the population is exposed.[87]

Not only would this be logistically feasible but it would provide information on which we could act. At the moment, we cannot take action to ban specific groups of pollutants suspected of being carcinogenic (phenoxy herbicides, organochlorines, etc.) At best, we can incriminate one or two individual chemicals which are then treated as scapegoats for the rest. Nor can we take action to prevent the release of poisons into our environment as a whole but only into certain parts of it, where the damage has been carefully documented by innumerable measurements, leaving us free to export the pollution to other areas where the effect of the pollutant is, and always will be, less well documented.

ELEMENT BOOKS LTD.
UNIT 25
LONGMEAD
SHAFTESBURY
DORSET SP7 8BR

Thank you for choosing this book.
If you would like to receive regular
information about Element titles,
please fill in this card.

Please tick the subjects that are of particular
interest to you

☐ PHILOSOPHY

☐ HEALTH & HEALING

☐ BUDDHISM, TAOISM

☐ WOMEN'S STUDIES

☐ EARTH MYSTERIES

☐ NEW SCIENCE

☐ CHRISTIANITY

☐ MYTHOLOGY

☐ YOGA

☐ ANCIENT WISDOM, ASTROLOGY, TAROT

☐ NATIVE AMERICAN

☐ HINDUISM

☐ QABALAH

☐ FICTION

☐ PSYCHOLOGY

☐ SUFISM, ISLAM

☐ WESTERN MYSTERY TRADITION

Other subjects of interest ....

....

Name ....

Address ....

....

....

On the basis of today's criteria it is possible for manufacturers to make out a case for the innocence of each one of the four million or so pollutants that they generate directly or indirectly, as a by-product of their activities, a case that can rarely be refuted on the basis of currently accepted scientific methodology. *Yet we know that between them, these pollutants are, among other things, causing the deaths of several million people a year from cancer.* Though we cannot prove that individual pollutants are contributing to this damage, their guilt *when seen as a group, is incontestable.* This principle not only applies to the study of how pollution affects natural systems but to the study of natural systems themselves, indeed to that of the biosphere as a whole.

Professors Jay Forrester, Denis Meadows and others have pointed out how the reductionist methodology of modern science does not enable one to understand the behaviour of natural systems. It must be remembered that natural systems are above all organizations which means that they are more than the sum of their parts, their identity and main characteristics being derived very largely *from the way in which these parts are organized.* This means that they cannot be understood simply by examining and measuring these parts individually and in isolation from each other, which is basically what our scientists are still trying to do, but only in the light of a general model reflecting not only their relationship to their own component parts but also to the larger systems of which they in turn are part. Such a model need not be quantitative. What we are interested in are the generalities not the particularities, the theoretical principles involved not just a mass of undigested quantitative data. Also it is not by measurement that we can determine what are these principles. In the scientific world of today, *measurement has largely replaced thought.* Thinking, in fact, has gone out of fashion. If we want to understand how the world works and how we are to adapt to it we must learn to think again and not with the aid of those clumsy machines called computers but with our brains which are infinitely more sophisticated pieces of equipment.

Let us then try to consider how we could examine pollution in its total biospheric context.

THEORETICAL CONSIDERATIONS

It took several thousand million years of evolution for the biosphere or world of living things, of which we are an integral part, to take on the shape industrial society found it in, and thereby provide an ideal habitat for humans and the myriads of other forms of life that compose it.

During the course of this evolution, as Commoner puts it,

> The chemical, physical and biological properties of the earth's surface gradually achieved a state of dynamic equilibrium, characterized by processes which link together the living and non-living constituents of the environment. Thus were formed the great elementary cycles which govern the movement of carbon, oxygen and nitrogen in the environment, each cycle being elaborately branched to form an intricate fabric of ecological interactions. In this dynamic balance, the chemical capabilities of living things are crucial, for they provide the driving force for the ecological cycles; it is the chemistry of photosynthesis in green plants, for example, which converts the sun's energy to food, fibre and fuel.[88]

The biosphere can function as a self-regulating natural system and maintain its basic structure, on which the very survival of its living components depend, only if the critical interrelationships between all its components—at all levels of organization, including that of the atom or the molecule—are maintained.

As Commoner further points out '. . . the chemical processes which are mediated by the biochemical system represent an exceedingly small fraction of the reactions that are *possible* among the chemical constituents of living cells. This principle explains the frequency with which synthetic substances that do not occur in natural biological systems . . . turn out to be toxic.'

Commoner illustrates this principle thus:

(a) Of the approximately one hundred chemical elements which occur in the materials of the earth's surface, less than twenty appear to participate in biochemical processes, although some of those which are excluded, such as mercury or lead, can in fact react quite readily with natural constituents.

(b) Although oxygen and nitrogen atoms are common in the organic compounds found in living systems, biochemical constituents which include chemical groupings in which nitrogen and oxygen atoms are linked to each other are very rare.

(c) Although the numerous organic compounds which occur in biochemical systems are readily chlorinated by appropriate artificial reactions, and the chloride ion is quite common in these systems, chlorinated derivatives are extremely rare in natural biochemical systems.

It is no coincidence that these chemicals are not found in living tissues. There is good reason for it. The organization that is the biosphere, has been able to evolve at the expense of eliminating possible reactions between these substances and living things. If any living systems once included them, then they have been eliminated by natural selection.

The consistent absence of a chemical constituent from natural biological systems is an extraordinarily meaningful fact. It can be regarded as prima facie evidence that, with a considerable probability, the substance may be incompatible with the successful operation of the elaborately evolved, exceedingly complex network of reactions which constitutes the biochemical systems of living things.

Furthermore, such theoretical considerations can be confirmed empirically.

Thus mercury is one of those eighty elements not essential for living processes. There is at least one good reason for this. Biochemical systems have evolved a system of enzymatic catalysis in which sulphur-containing groups play a crucial role. These react with mercury introduced into a living system, and enzymes are inactivated, often with fatal results.

There is also a good reason why synthetic nitroso compounds in which nitrogen and oxygen atoms are linked do not occur either in living tissue. They appear to interfere with the reactions involved in the orderly development of cells, and give rise to cancer and mutations.

There is also a good reason why synthetic organochlorine compounds such as DDT and PCBs are excluded from living tissue. They are often very toxic or produce long-term damage such as cancer.

How does a living system succeed in excluding unwanted chemicals? The answer is that either these chemicals are not present in its environment in that form which would permit them to interfere with it, or the system develops subtle homeostatic mechanisms for maintaining low levels within it, even if the levels outside are higher. These mechanisms, however, have developed via the evolutionary process—hence very slowly. They can only deal with chemicals found in that form and at that level to which the system was exposed during its evolutionary experience. In general the more the environment changes as a result of human activities, the less does it resemble that in which we evolved, and the less efficiently can our normal behavioural mechanisms enable us to adapt to it. Thus, while the human liver is capable of detoxifying those chemicals that it has learnt to detoxify over millions of years of human evolution it is incapable of detoxifying chemicals to which it has not been exposed during this period.

It is these considerations which led Professor Stephen Boyden of the Australian National University to formulate his principle of phylogenetic maladjustment.[89] He pointed out that since the evolutionary process is adaptive, it must be when subjected to that environment with which we have co-evolved that our biological needs are best satisfied. This means that any modification of our environment causing it to divert from that to which we have been adapted by our evolution must lead to phylogenetic or evolutionary maladjustments and the greater this diversion the greater these maladjustments must be.

BANNING POLLUTANTS

From the preceding analysis it should be clear that to avoid the rapid deterioration of the biosphere and the corresponding reduction in its capacity to support complex forms of life such as humans, there is no alternative but to reduce very considerably our environment's total pollution-load.

This cannot be done by examining individual pollutants by the reductionist method in controlled laboratory conditions, but only on the basis of a model that takes into account both theoretical and empirical factors, in terms of which the probability of the harmfulness of different chemicals can alone be established. The degree of probability required must vary with the extent of the damage that a specific pollutant is suspected of causing. For instance, if it could be implicated in causing cancer or mutations or in possible climatic changes, then clearly *the slightest probability of its guilt must be regarded as sufficient to warrant its removal from the market.*

The chemicals that must first be withdrawn are largely those which have been introduced in the last thirty years—during which time, as Commoner has pointed out so convincingly, pollution levels have escalated in the US by between 200 and 1,000 per cent—totally out of proportion to the economic growth registered during this period and even more so with any possible benefits we might have derived from their use.[90]

Foremost among these chemicals are the synthetic organics which must include the synthetic nitroso and organochlorine compounds mentioned by Commoner. There are some nine thousand of them mainly used as plasticizers, aerosol propellants, refrigerants, pesticides and herbicides. According to Epstein 'Very few, if any of these compounds are without toxic effects, either because of their own chemical properties, or because of chemicals discharged to the environment during their manufacture, or because of breakdown products, or because of some potentiating, synergistic effect when they come into contact with other chemicals.'[91] Yet as Saffiotti points out 'only a small proportion of these substances are exhaustively tested against the possible hazards contingent upon wide dispersion in the environment.'[92]

These are only the most obvious ones, the list of all the toxic chemicals that we are releasing in an almost uncontrolled manner into our environment would be a much longer one; it would include the several thousand chemicals we add to our food during processing, few of which according to Ross Hume Hall 'have received more than a cursory examination', or have been rigorously tested for their ability to cause 'birth defects, heart attacks, cancer and behavioural abnormalities.'[93]

It would include nitrites that are used so extensively as food preservatives, and nitrogen fertilizers, whose massive use is leading to an equally massive increase in the nitrate content of our drinking water. But banning the use of such substances would not be sufficient. Drastic reductions would be required in emissions of SO_2, NO_x and CO_2 to the atmosphere.

It is doubtful if pollution levels in our society could be reduced by any other means than by deliberately reducing the level of our industrial activities. This would mean giving up the goal of 'material progress' and setting out as Robert Prescott Allen and I proposed in *A Blueprint for Survival* to create a totally different non-industrial society, one in which economic and political activities were carried out on a very much smaller scale.

WILL THERE BE ANY REAL ATTEMPT TO CONTROL POLLUTION?

On the basis of past experience we know that, unless the Green Party were to form a government, such a programme would never be adopted. Things are done in our industrial society to satisfy three sets of requirements, those of our industrialists who want higher profits, those of our trade unionists who want more jobs at an ever higher rate of pay, and those of our politicians who want more votes. Profits, jobs and votes, are seen as best obtained by maximizing economic activities and hence pollution. We can thereby predict that the acceptable levels for different pollutants will remain as high as public opinion will allow polluters to keep them, that dangerous substances will not be banned unless they give rise to immediate visible large-scale catastrophies such as those which occurred at Minamata, at Seveso and, since this chapter was written, at Chernobyl, and even then they will probably only be banned locally and for a short period. The public's memory is notoriously short.

To justify its inaction, our government will make use of every subterfuge to con the public into believing that pollution is under control. Thus, it will persuade successive committees of learned experts to fix unduly high permissible levels for the different pollutants in our environment. Measurements will continue to be conducted and interpreted in such a way as to allay public fears. Additive and synergistic effects and the effect of decay products and impurities will continue to be disregarded. The accent will remain on short-term toxicological effects while long-term carcinogenic and mutagenic effects will continue to be played down. The absence of hard 'scientific evidence' to prove the harmfulness of particular chemicals will remain an obvious excuse for inaction—and as little money as possible will be spent on obtaining this evidence.

Lack of funds and the adverse effect on our standard of living of spending too much money on pollution controls will be another. When action is taken it will be, as today, largely for cosmetic purposes. As Professor Kreith writes

> The government is more likely to be concerned with ameliorating the feelings of the public, of alleviating those factors that are visible and are the source of public controversy. For instance, when requests arise for cleaning stacks, industry may remove the steam which is visible, but disregard the more dangerous sulphur dioxide, which is invisible but much more difficult to remove from the exhaust,[94]

just as our government has done so far in the UK.

When our government is forced by public opinion to pass legislation designed to prevent further environmental contamination, one can predict in advance that such legislation will either be so emasculated that it will have little effect or else that it will never be implemented as has largely been the case with the 1974 Control of Pollution Act.

WHAT OF THE FUTURE?

For all these reasons, one can assume that the vast bulk of the pollution generated by our industrial activities will find its way into our environment which means that total pollution emissions to the environment will, to all intents and purposes, reflect closely the level of industrial activity.

This conclusion is implicit in most of the serious forecasts of pollution trends in Europe. The Economic Commission for Europe points out, for instance, that in spite of all measures taken to control the release of waste products of all sorts into the European environment, it is continuing to increase at a rate of about 5 per cent per annum, while the quantities of inorganic waste released into the environment worldwide will continue to double every ten to twelve years.[95]

In another, little publicized, OECD report it is admitted that the OECD area is rapidly reaching the point where it must choose between industrial expansion and clean air. The report predicted that emissions of nitrogen oxides and sulphur dioxide from the burning of fossil fuels would go on increasing unless there was a reduction in fuel consumption and by implication of economic activity.[96]

Already, in one year, it appears the waste the European community has released into the environment includes 90 million tonnes of household refuse, 115 million tonnes of industrial waste, 950 million tonnes of agricultural waste, 200 million tonnes of sewage sludge and 150 million tonnes of waste from extractive industries.

The physical problem of disposing of such massive quantities of waste products is in itself a major one, and the danger to public health is already, the Commission admits, serious. Yet by the end of the century, if economic activity continues at the present rate, the quantities will have quadrupled—with wastes accumulating on the land, in rivers and waterways and in the atmosphere, and often too in biological organisms including human ones with inevitable detrimental effects on health.

Pollution by radioactive materials must also increase in the same way. Already, as Dr Spearing points out,

> merely the 'low level' releases to the environment, currently occurring, contain long-lived radio-isotopes which are being discharged *at a rate exceeding the rate at which their radioactivity is decaying*. In consequence there is a gradual and insidious build-up of environmental radioactivity, and there is a very real risk of irreversible contamination of our planet to a degree that will impose a severe burden of human suffering on future generations quite possibly to the end of the story of human life on earth.[97]

As Sir Brian Flowers warns

> By the year 2000, a world nuclear power programme would have generated such large quantities of fission products (and actinides) that even if they were dispersed uniformly in the vast bulk of the oceans, the resulting concentration would be within one or two orders of magnitude of the maximum permissible concentration for drinking water. This would

not be satisfactory because of the many food chains that would concentrate the radioisotopes and return them to man.[98]

With regard to marine pollution in general, one of the world's foremost oceanographers, Dr Edward Goldberg, writes

> Our concern is the haunting possibility that levels of a toxic material can rise so high that exposure of organisms to such materials in the open ocean, as well as in the coastal ocean, may result in widespread mortality or disease . . . If these substances mix with the deep ocean, they will be transferred within a decade to zones below the mixed layer, where they may remain for thousands of years . . .

He concludes that we may leave future generations 'the legacy of a poisonous ocean'.[99]

Another consequence of the increased contamination of our planet must be the continued incidence of cancer. Already more than 25 per cent (fifty-one millions) of the two hundred million people living in the US will get cancer. Thirty-four millions will die of it. As Epstein points out 'most of the people dying today are over forty or fifty years old and were thereby brought up in that period that preceded the general contamination of our environment by most of the known carcinogens in general use today. We can thereby expect that when today's children reach the age of forty or fifty, the cancer rate will be very much higher.'[100] Frank Rauschel when he was Director of the National Cancer Institute agreed with this thesis. 'Given today's environment' he wrote, 'we are living with a time-bomb that's going to explode in twenty or thirty years from now in the form of even more persons being stricken with cancer.'[101]

Indeed at the rate at which the cancer rate is increasing today, it is only a matter of a few decades before this dreaded disease becomes generalized among the populations of industrial countries—a truly nightmarish prospect. It is now generalized among fish populations in highly polluted US East coast rivers.

However, perhaps one of the most dramatic consequences of present pollution trends must be changing climatic patterns. Professor Flohn at the Second International Conference on the Environmental Future, went so far as to state that 'a global climatic catastrophe is unavoidable, if we continue to use energy at the current rate', a conclusion that was also that of other eminent climatologists present. Indeed it is difficult to see how such a conclusion can be avoided if one accepts with Flohn that we are already 'on the fringe when man-made changes' to the chemical composition of the atmosphere 'are at the same level as natural ones'—and are, what is more, still increasing.[102]

WHAT HOPE IS THERE?

In the introduction of the Fifth Report of the Royal Commission on Environmental Pollution, Mr Crossland, who was then Minister of the Environment, congratulates its authors for showing that there was no substance to the predictions by environmentalists that our industrial activities were causing irreversible damage to our environment.

In the same report, its principal author, Sir Brian Flowers, concludes that pollution could never by itself limit economic growth. These statements, which reflect official opinion in this and other countries as well, could not be further from the truth. Indeed if global environmental pollution were to increase at the current rate for more than a few decades, economic activities like all other human activities, would be dramatically curtailed by the mere fact that our planet would have ceased to provide a suitable habitat for complex forms of life such as human and the other higher mammals.

In reality of course such a situation is unlikely to occur. Over the next decades our polluting activities are likely to diminish rather than increase. This, however, is not going to be because of any intelligent decisions taken either by our industrialists, or our politicians, but simply *because world conditions are becoming ever less propitious to the industrial process*. It is, in fact, global economic catastrophe that is likely to provide the only effective method of pollution control.

Notes:

1. Lord Zuckerman, The Environment. *This Month*, London, 1972.
2. Nicholas Wade, *Science*, 13th Feb. 1976.

3. The PCB Problem, New Canadian Report Pollution. *Environmental Bulletin*, Oct. 1976.
4. John Nisbett, Balancing the Costs of Cancer, *Technology Review*, Jan. 1976.
5. Lawrence McGinty, *New Scientist*, 14th July, 1977.
6. Ross Hume Hall, *Food for Nought*, New York, Doubleday, 1974.
7. Anthony Tucker, *The Guardian*, 7th Oct. 1977.
8. *Science News*, Vol.102, September 1970.
9. *New Scientist*, 15th Jan. 1976.
10. *Environmental Policy*, No.82, Nov. 26th 1978.
11. Peter Schmidt, *Alsdorf*, 1969.
12. Robert Walgate, *Nature*, Vol.280, 5th July 1979.
13. *Nature*, Vol.280, 5th July 1979.
14. *New Scientist* 22nd July, 1976.
15. Blind Man's Bluff, *Ecologist Quarterly*, Spring 1978.
16. Samual Epstein, *Testimony on the Delaney Amendment and on Mechanisms for reducing Constraints in the Regulatory Process, in general and as applied to Food Additives in particular*, US Senate Hearings before the Select Committee on Nutrition and Human Needs, 20th Sept, 1972.
17. Anita Johnson, *Environment*, April, 1979.
18. Neville Grant, Mercury in Man, *Environment*, May 1971.
19. Anthony Tucker, *The Guardian*, 7th Oct. 1977.
20. Derek Bryce-Smith, text of unpublished lecture.
21. *Nature*, Vol.276, 16th November, 1978.
22. Prof. Gordon Athersley, Proof of Evidence, Windscale Inquiry, Sept. 1977.
23. *Chemical and Engineering News*, 19th Jan, 1976.
24. Jack Schubert, The Programme to Abolish Harmful Chemicals, *Ambio*, 1972.
25. Neville Grant, op.cit.
26. Anthony Tucker, *The Toxic Metals*. London, Pan/Ballantyne, 1972.
27. E. Sturgess, *The Times* quoted in *The Ecologist*, Vol.3, No.9, Sept. 1973.
28. Colin Norman and Chris Sherwell, Treading Softly on the Ozone Layer, *Nature*, Vol.263, 23rd Sept. 1976.
29. Angela Singer, *The Guardian*, 22nd April, 1978.
30. Michael Waldichuk, *The Assessment of sub-lethal effects of pollution in the Sea. Review of the problems*. Paper prepared for the Royal Society Meeting on 24th/25th May, 1978.
31. H. Rosenthal and D. R. Alderdice, 1976. Sub-lethal effects of environmental stressors, natural and pollutional, on marine fish eggs and larvae. *J. Fish, Res. Board Can.*, 33, 2047-2065 (quoted by Waldichuck), op.cit.
32. E. M. Donaldson and H. M. Dye, 1975, Corticosteroid concentrations in sockeye salmon (*Oncorhynchus nerka*) exposed to low concentrations of copper. *J. Fish. Res. Board Can.* 32, 533-539 (quoted by Waldichuck), op.cit.
33. J. M. Anderson, *Assessment of the effects of pollutants on physiology and behaviour*, Proc. Royal Soc. London, B177, 307-320, 1971, (quoted by Waldichuck), op.cit.
34. H. Kleerekoper, 1976, Effects of sub-lethal concentrations of pollutants on the behaviour of fish. *J. Fish Res. Board Can.* 32, 2036-2039 (quoted by Waldichuk), op.cit.

35. P. E. Lindahl and E. Schwanbom, 1971, Rotary-flow technique as a means of detecting sub-lethal poisoning in fish populations *Oikos*, 22, 354-357 (quoted by Waldichuk), op.cit.
36. H. Rosenthal, op.cit.
37. Anthony Tucker, 1972, op.cit.
38. S. G. Rainsford, quoted by Anthony Tucker, 1972, op.cit.
39. Anthony Tucker, 1972, op.cit.
40. I. M. Dale, M. S. Baxter, *et al.* Mercury in Seabirds, *Marine Pollution Bulletin*, Vol.4, No.5, May, 1977.
41. Carroll Wilson, *et al.* A Study of Critical Environmental Problems (SCEP) Man's impact on the Global Environment, Cambridge, Mass. M.I.T. Press, 1971.
42. Irving Selikoff, *Nature*, December, 1978.
43. John D. Walker and Peter R. Caldwell, Deadly Mercury and Bacterial Interaction, *Environmental Science and Technology*, Vol.10, No.12, Nov. 1976.
44. Jack Schubert, op.cit.
45. Carroll Wilson, op.cit.
46. O. Linden, Acute effects of 081 and Oil Dispersant Mixtures on Larvae and Baltic Herring, *Ambio*, Vol.7, No.2, Feb. 1976.
47. Janice Crossland and Virginia Brodine, Drinking Water, *Environment*, April, 1973.
48. Samuel Epstein, The Delaney Amendment, *The Ecologist*, Vol.3, No.1, 1973.
49. Jack Schubert, op.cit.
50. Bryn Bridges, *The Ecologist*, June 1972.
51. Jean Marx, *Science*, Vol.201, 11th Aug. 1978.
52. Samuel Epstein, op.cit.
53. Anthony Tucker, 1972, op.cit.
54. V. T. Bowen, Proof of evidence at the Windscale Inquiry.
55. Peter Bunyard, Gearing up to the Plutonium Economy, *The Ecologist*, Vol.6, No.10, 1976.
56. Environmental Directorate Report OECD, 1975.
57. Colin Norman and Chris Sherwell, op.cit.
58. When to Ban Nasty Chemicals, *New Scientist*, 8th April 1976.
59. Charles Wurster, The effects of pesticides on the Environmental Future *in* Nicholas Polunin (ed) *The Environmental Future*, London, Macmillan, 1973.
60. Abel Wolman, Ecological dilemma, *Science*, No.193, 1976.
61. Rene Dubos, *Pollution*, Publication of the National Association of Biology Teachers, 1973.
62. Anthony Tucker, The Poison Cloud over Civilization's Future, *The Guardian*, 5th August, 1976.
63. Janice Crossland, Drinking Water, *Environment*, Vol.15, No.3, April 1973.
64. M. Fielding and R. F. Packman, Organic Compounds in Drinking Water and Public Health, *The Ecologist Quarterly*, No.3, Summer 1978.
65. Pollution, *Environment News Bulletin*, May, 1976.
66. *Nature*, Vol.261, No.409, June 1976.
67. *Archives of Toxicology*, Vol.29, No.287, 1975.
68. Jon Tinker, Treating Smog with Air Freshener, *New Scientist*, Sept, 1976.
69. Environmental Directorate OECD, 1976.
70. J. Sontheimer, *The Ecologist*, Vol.3, No.9, September 1973.

71. Spilling the Beans, *The Economist*, 7th Feb. 1976.
72. W. C. Chalmers, Evidence presented at the Windscale Inquiry, day 48, p.32, 1977.
73. Charles Wakstein, British Nuclear Fuel's safety record, *The Ecologist*, Vol.7, No.6, Oct 1977.
74. Sir John Hill, quoted by Beryl Kemp in Nuclear Alert, *Conservation News* Jan. 1977.
75. J. T. Edsall, *Environmental Conservation*, Vol.1, No.1, 1979.
76. Sir Richard Doll, *Nature*, 265, Feb 17th 1977.
77. Samuel Epstein, op.cit.
78. Drug Safety and the Expert, *New Scientist*, 2nd Sept. 1976.
79. Samuel Epstein, op.cit.
80. John E. Blodgett, Pesticides, Regulation of an evolving technolgy *in* (Samuel Epstein and Richard Grundy, eds), *Consumer Health and Product Hazards/Cosmetics and Drugs, Pesticides, Food Additives*, Vol.2, Cambridge, Mass. The MIT Press, 1974.
81. Rationalizing the Hunt for Cancer Causing Chemicals, *The New Scientist*, April, 1976.
82. *Not Man Apart*, April 1976.
83. D. Saffiotti quoted in Samuel Epstein and Richard Grundy eds. op.cit.
84. Epstein and Grundy eds. ibid.
85. D. Saffiotti quoted by Epstein and Grundy eds. ibid.
86. Alvin Weinberg, quoted by Epstein and Grundy eds. ibid.
87. Jack Schubert, op.cit.
88. Barry Commoner, Foreword to Epstein and Grundy eds. op.cit.
89. Stephen Boyden, Evolution and Health, *The Ecologist*, Vol.3, No.8, August 1973.
90. Barry Commoner, *The Closing Circle*, Jonathan Cape, London 1971.
91. Samuel Epstein, op.cit.
92. D. Saffiotti, quoted by Epstein, op.cit.
93. Ross Hume Hall, op.cit.
94. Frank Kreith, Lack of Impact, *Environment*, Vol.15, No.1, Jan/Feb 1973.
95. Pollution, *Environment News Bulletin*, 26th Sept. 1977.
96. *The Ecologist*, Vol.3, No.1, 1973.
97. Dr. J. K Spearing, Proof of Evidence, Windscale Inquiry, 1977.
98. Sir Brian Flowers in *6th Report of the Royal Commission on Environmental Pollution*, HMSO, London.
99. Edward Goldberg, *The Health of Oceans*, Unesco Feature, 1975.
100. Samuel Epstein, The Biological and Economic Basis of Cancer, *Technology Review*, Vol. 78, No.8, July/Aug. 1976.
101. Frank Rauschel, *US News and World Report*, 1976.
102. Professor Flohn quoted by Edward Goldsmith in The Reykjavik Conference, *The Ecologist*, Vol.7, No.6, 1977.

CHAPTER SIX

The Ecology of War

'But if we don't export our food, how can we obtain all the armaments to maintain our national security?'

The Ecology
of War

NEVER BEFORE HAVE so many governments throughout the world committed themselves so piously and so persistently to the ideal of world peace—yet never before has this world been ravaged by so many wars and on so vast a scale.

This seeming paradox, our politicians would undoubtedly explain in terms of some technicality, such as a shortage of funds with which to implement their peace-making strategies. None, perhaps, would even conceive the possibility that it was the strategies themselves that were at fault. Yet this is the only conclusion that is reconcilable with our knowledge of the problems involved, both empirical and theoretical.

Let us consider our present attitudes to war and methods of controlling it. First, it is held to be irrational, since it reduces material wealth by killing people and destroying property. People on the other hand are regarded as rational. It must follow that war can only be the result of misunderstandings which can surely be dispelled if our politicians be allowed to meet for heart-to-heart talks. Hence the extraordinary faith people seem to attach to summit conferences as a means of averting wars, an illusion fully exploited by several of our recent prime ministers for whom the setting up of such conferences was one of their major goals, on the basis of whose achievement they persuaded the electorate to judge the success of their government.

If negotiation is the answer, then it is clearly important to do so we are told 'from a position of strength', hence the arms race, or as it is euphemistically called mutual deterrence. With the passing of the Strategic Arms Limitation Treaty (SALT) mutual deterrence has been consecrated as the principal means of controlling wars. The agreement attempts to control the arms race by applying quantitative limitations on arms build-up. However, its effectiveness is seriously compromised by the fact that it makes no mention of qualitative controls. As Frank Barnaby points out, the nuclear arms race, in any case, was about to become a race for quality rather than quantity.[1] This means that a new premium has now been placed on research and development into ever more lethal weaponry. It is not surprising that the US budget for research and development in this field is likely to increase beyond the present eight billion a year (1974), about the same as is spent by the USSR—a sum greater than any other country devotes to its entire military budget. It seems quite extraordinary that we should have created a situation in which a world war of unprecedented horror can only be averted by spending such gigantic sums on ever more sinister instruments of death.

Mutual deterrence is the industrial solution par excellence. It involves for ever expanding the world's industrial machine as, if nothing else, a means of providing armaments for preventing war. Yet as I shall attempt to show, it is this process of industrialization which is causing the maladjustments which must both increase the probability as well as the destructiveness of war. Never, in fact, was a policy so evidently self-defeating.

If we want to control warfare we must first understand why it occurs, which must mean examining aggression, i.e. that feature of animal behaviour which, in the case of human animals at least, gives rise to war.

One must first distinguish between two fundamentally different types of aggression; that displayed by a predator towards its prey and that displayed by an animal towards a rival animal of the same species. First of all the goal is different. In the first case it is to seize the prey and eat it. In the second it is to establish a position in the hierarchy, which may enable a male to claim rights over a particular group of females or a particular territory. Its object is not to kill but to subdue a rival so that he accepts a lesser position in the hierarchy.

Since the goal is different, the behaviour that will achieve it must also be different. Thus a predator will approach its prey as stealthily and noiselessly as possible. When accosting a rival, however, an animal will behave in precisely the opposite manner. The lion will roar, the dog will bark, the peacock will swell himself up so as to intimidate his rival with his size and splendour. The effect can only be to frighten away a rival, the last thing a predator wishes to do to its prey. These two types of aggression are kept distinct in the animal world.

Significantly, warfare among human beings was conducted very much along the latter lines until a short time ago. Soldiers would wear the most flamboyant uniforms and go into battle to the blare of trumpets and the beating of drums. In present-day warfare, however, there is indeed little ritual left. Soldiers are dressed in camouflage so that they can creep up on their enemies unobserved, like predators on their prey. Indeed, the goal of today's warfare is not to frighten but to destroy the enemy, so as to acquire for one's own use its land and other resources. In this way modern warfare has diverged very radically from that carried out among traditional societies. But then, our society is highly aberrant, which means that to understand warfare, we must begin by examining it as it occurs among non-human animals and among tribal societies—which are the normal units of social organizations.

If we do so, the first fact that can be established is that aggression is a fact of life. As Eibl Eibesfeldt writes, 'fighting between members of the same species is almost universal among vertebrates from fish to man.'[2] This is no hazard. Animals are simply designed that way, which is a fact that is confirmed by physiological studies of the neural and hormonal processes involved.[3] Researchers have even succeeded in inducing fighting behaviour among birds and mammals by stimulating specific areas of the brain with electric currents.

Humans are no exception to the rule, though many of us would like to think we are. Some anthropologists like Derek Freeman consider that, if anything, humans are more aggressive than other animals. According to him, 'the extreme nature of human destructiveness and cruelty is one of the principal characteristics which marks off man behaviourally from other animals.'[4] This point has been cogently expressed by the biologist Adolf Portman:

> When terrible things, cruelties hardly conceivable, occur among men, many speak thoughtlessly of 'brutality', of bestialism or a return to animal levels. As if there were animals which inflict on their own kind what men can do to men. Just at this point, the zoologist has to draw a clear line, these evil horrible things are no animal survival that happened to be carried along in the imperceptible transition from animal to man, this evil belongs entirely on this side of the dividing line, it is purely human.[5]

Whether humans living in their natural habitat are as unpleasant animals as Freeman makes them out to be, is very debatable. Freeman may have been deceived by the fact that most societies we know about are, to a large extent, aberrant.

The Comanches, for instance, were regarded as particularly bloodthirsty warriors, but they were not always so. They came to adopt that way of life only after they had come in contact with the white invaders who, among other things, provided them with horses. To quote Kardiner, 'there were no aggressive war patterns in the older culture and little intertribal fighting. Though there was a little fighting with warlike neighbours there were no special honour aspects to war.'[6]

The same can also be said of the Zulus. The career of Chaka, the founder of the Zulu Empire, was considerably affected by contact with the Boers whose activities gave rise to the freak situation he so fully exploited.

As I shall show later, aggression between human societies, living in an environment to which they have been adapted phylogenetically and culturally, takes a far less destructive form.

The fact that aggression must be a feature of social behaviour among non-human and human animals can be deduced from the fact that it fulfils important functions, on whose fulfilment, societies depend for their very survival. Thus Eibl Eibesfeldt regards aggression as serving 'the important function of spacing out individuals or groups in the area they occupy. This thereby secures for each the minimum territory required to support its existence, prevents overcrowding and promotes distribution of the species.'[7]

Aggression among primitive societies, in particular among hunter gatherers, leads to fission of the group into rival factions which part company and develop on their own. This prevents society from getting too big. This is particularly important as the bonds holding together a society, as Malinowski was perhaps the first to point out, are derived from those which serve to hold a family together and cannot be extended to hold together too big a social unit. When the maximum size is reached, the unit must break up, alternatively it disintegrates into a mass society incapable of adaptive behaviour. It is in this light that Margaret Mead interprets the fission of social groups among the Maori. 'The disintegrative potentiality in the Maori form of organisation was this factor of size: when the society grew too large it was not possible to maintain their complete identification, and a subgroup would split off and become another autonomous, closed, co-operative unit, competing without and co-operating within.'[8]

In this way, a society remains a cohesive unit characterized more by co-operation than by competition, and bickering and fighting is reduced to a minimum.

COMPETITION AND CO-OPERATION

Aggression is best regarded as a form of competition, which, as we shall see, takes a slightly different form as we move up from one level of organization to the next. At each step, it plays a bigger role, and can become more violent, i.e. can resemble more what we call aggression, without thereby bringing about social collapse.

Undifferentiated individuals competing for the same ecological niche cannot co-operate in any way. They can only compete with each other. It is only when, as a result of competition, they have been forced to specialize in such a way that each one learns to exploit a different sub-niche, that co-operation becomes possible and the competing individuals are transformed into a viable social unit. It is only by competition therefore that conditions are established in which co-operation can occur.

Competition does not establish just any type of organization, but that which best satisfies environmental requirements, i.e. that which is most adaptive. Thus, in a social system that earns its livelihood by hunting, the position of an individual in the hierarchy will depend on hunting ability. In a society in which the main activity is warfare, war-like qualities will be determinant.

It is important to note that the basis of a hierarchical structure will change in accordance with its adaptiveness. Thus any important change in environmental conditions will call for a modification in a system's behaviour pattern which can only be ensured by a reorganization of its hierarchical structure in such a way that a premium is placed on the new qualities which the society must display in order to adapt to the new conditions.

The influence of an individual will depend on his position in the hierarchy and so will his command over the best territory, the most desirable females, etc. In times of serious shortage this means that he is less likely to succumb than are those lower down in the hierarchy. Seen over a longer time scale, it is the individuals at the top of the hierarchy who are most likely to transmit their characteristics to their progeny.

Competition also serves the purpose of eliminating deviants who, for various reasons, do not fit into the social structure. In a traditional society this is probably a minor function of competition as, on the whole, most cultural patterns provide socially acceptable outlets for predictable deviant forms.

In the human family, the relative roles of the father, the mother, the children, the grandmother, etc. have to a certain extent been determined biologically, while cultural tradition, transmitted during socialization will complete the differentiation process. In such conditions, there is little need for much competition. Aggression has some role to play in sexual behaviour in many animals, probably also among humans in the initial stages of a relationship. Its role may well be to establish the basis of what then becomes a co-operative relationship.

As we move from the family to the community, so does competition tend to increase. Competition is in fact necessary to establish an individual's status when this has not been established by inheritance. Societies vary according to their degree of co-operation and competition. *The competitive society will have the advantage of being able to modify at more frequent intervals the basis of its hierarchical structure. This will enable it to adapt to a rapidly changing environment.* It does not appear to have any advantage however in a relatively static environment.

Since aggression is a basic feature of human behaviour, the idea of eliminating it altogether, to bring about the universal human family is naive, while efforts to achieve this can only be counterproductive. In traditional societies, however, a number of strategies are exploited to reduce its destructiveness to the very minimum. One such strategy, as we have seen, is fission. Significantly when population pressure increases, fission is no longer practicable. On the contrary, the distance between social groups is increasingly reduced, and warfare becomes correspondingly more destructive. Thus among the Nuer in the southern Sudan whose territory has shrunk considerably as a result of increased population pressure, tribal disputes according to McDermott lead to considerably more casualties than was once the case.[9]

RITUALIZATION

One of the most important means of reducing the destructiveness of aggression, as already noted, is to ritualize it. This means providing it with an outlet at the minimum cost in terms of physical damage and loss of life.

The very principle of ritualization is alien to industrial society which places a premium on achieving the maximum results with the minimum human effort. Ritualization ensures the opposite: *the achievement of the minimum results with the maximum human effort. Contrary to what we have all been taught, it is often the latter goal not the former which is most adaptive.*[10]

The ritualization of aggression is a feature of the behaviour of all animals including humans . That it should be so widespread is not surprising. As Lorenz points out,

A particularly successful solution is often found by the different branches independently of one another. Insects, fishes, birds and bats have 'invented' wings, squids, fishes, ichthyosaurs and whales the torpedo form. It is not surprising that fight-preventing behaviour mechanisms based on ritualised redirection of attack occur in analogous developments in many different animals.[11]

Among non-human animals, as already mentioned, inter-species conflict is simply not designed to lead to death or mutilation. It is more than anything else a ritual and is conducted according to a set of rules designed above all to prevent death or mutilation from occurring. Thus rattlesnakes, capable of killing each other with a single bite, never in fact bite a rival. Their conflict is a strange ritual resembling Indian wrestling. 'The successful snake pins the loser for a moment with the weight of his body and then lets him escape.'[12] The oryx antelope who is equipped with horns capable of putting a lion to flight does not use his horns as daggers in an interspecific fight. As Eibl Eibesfeldt writes, 'a hornless bull observed by Walthe carried out the full ritual of combat as if he still had horns. He struck at his opponent's horns and missed by the precise distance at which his non-existent horns would have made contact. Equally remarkable, his opponent acted as though his horns were in place and responded to his imaginary blows.'

Among birds and mammals conflict usually ends when the weaker contestant makes a recognized gesture of submission. Such appeasement gestures often consist in offering a vulnerable part of the body to the victor. Desmond Morris uses a term 'remotivating displays' to refer to 'actions which arouse in the attacker a non-aggressive tendency which then competes with and inhibits the aggressiveness that is present.'[13] Two common examples of this that occur in many species, are pseudo-infantile displays and pseudo-sexual displays. In these instances the submissive animal performs either juvenile or sexual patterns which arouse parental or sexual responses in the aggressor and in this way stop the attack. This triggers off a peace response on the part of the victor. In this way conflicts rarely end in serious mutilation or death.

Very much the same is true of aggression in human societies. Among hunter-gatherers such as the Bushmen or the Australian Aborigines when still living in conditions approaching those to which humans as a species have been adapted by evolution, warfare appears to be limited to border skirmishes giving rise to very few casualties.

This is illustrated by an incident recounted by Love.[14] In 1837 a party of men led by Sir George Gray, one of the first Englishmen to land in Australia, was assailed by a group of armed Aborigines who threw spears at them and gravely wounded Gray himself: as a result Gray had to fire at them and shot and killed one man. The others immediately dispersed but shortly came back to collect the man they had left behind who was wounded and subsequently died. It did not seem to have occurred to them that these foreigners might shoot at them again. We now know that it is the normal procedure in Australian aboriginal warfare for hostilities to cease at the death of one man, which provides the explanation for the otherwise inexplicable behaviour of the Aborigines.

Among the Maori, leadership was all important and hostilities came to an end once the leader of one of the rival groups was put out of action. Vayda writes: 'Even when on the verge of victory an attacking force might withdraw because of the loss of a leader.'[15]

Even among the most warlike tribes, the object of war was seldom to kill or to conquer territory which, in any case, they usually regarded as inhabited by hostile spirits. The tendency, on the contrary, was to stick to their own territory which was inhabited by the friendly spirits of their ancestors. This is even true of the specially warlike Jivaro Indians who, as Karston points out, 'inhabit endless virgin forests, where they can make new settlements almost anywhere and have no need of conquering the territory of other tribes.'[16]

Rather than kill or conquer territory the aim of such warlike tribes was to achieve certain largely ritualized goals. In the case of the Crow Indians, who were among the most warlike of American tribes, the first of such goals was to cut loose and steal a horse picketed within a hostile camp. The second was to take an enemy's bow or gun in a hand-to-hand encounter. The third was to touch an enemy with a weapon or even the bare hand. The fourth was to lead a victorious war expedition.[17]

In the middle of the nineteenth century, the Blackfoot Indians were the most warlike and the most feared of the Indian tribes in the valley of the Saskatchewan and the upper Missouri. Yet, as Ewers points out,

> Blackfoot warfare was aimed at neither the systematic extermination of enemy tribes nor the acquistion of their territory. It was not organised and directed by a central military authority nor was it prosecuted by large, disciplined armies. Rather, Blackfoot warfare was carried on primarily by numerous small parties of volunteers who banded together to capture horses from enemy tribes . . . While the killing of enemy tribesmen and the taking of scalps were not major objectives of these raids.[18]

Ewers describes in greater detail what were the most prestigious feats of war among the Blackfoot Indians.

> They graded war honours on the basis of the degree of courage displayed in winning them. They recognised that a man might scalp an enemy who had been killed by another and that a man might kill an enemy from a considerable distance with bullet or arrow. Their term for war honour, 'namach-kani' means literally 'a gun taken'. The capture of an enemy's gun ranked as the highest war honour. The capture of a bow, shield, war bonnet, war shirt, or ceremonial pipe was also a coup of high rank. The taking of a scalp ranked below these deeds, but ahead of the capture of a horse from the enemy. The capture of a horse was too common an accomplishment to receive a higher rating.

Another form of ritualization is the substitution of a match or tournament between two or more champions for a conflict between two armies. The contest between David and Goliath is an obvious example. This strategy was resorted to a great deal during the Middle Ages in Europe. The modern equivalent is, of course, sport, though this was also resorted to among tribal societies. Lacrosse, for instance, was used as a means of ritualizing conflict among the Creek Indians.

The transition from war to sport on the small island of Truk in Micronesia is traced by Murdock. This island was occupied successively by the Japanese, the Germans, and the Americans. The Germans abolished traditional warfare and the Americans introduced baseball as a means of replacing it. As Murdock writes,

The natives don't play baseball, they wage it. An inter-island game is serious business. The players practise almost daily and observe all the sexual and other taboos which used to precede war. For several days before a game, for example, they sleep apart from their wives in the men's clubhouse. The women and children form groups around the playing field, singing songs and executing magical dances designed to discommode the foe. Special baseball songs constitute one of the principal forms of native music.[19]

In our industrial society, football plays an important part too in ritualizing latent aggression between various groups. This is so, for instance, in Glasgow where a match between the two principal teams, one of which represents the local Protestants, the other the immigrant Irish Catholics, arouses tremendous enthusiasm on both sides.

The most sophisticated example of the ritualization of aggression in a modern society is the Palio in Siena, a horse race around the Piazza del Campo, the city's main square. This city of about 70,000 people has probably since the Middle Ages been divided into a number of 'Contrades' of which there are now seventeen. These were originally military associations whose function it was to raise troops to serve in the continual wars of the Sienese Republic against its neighbours. The Contrades, most of which are named after animals, such as 'Oca' the goose, 'Istrice' the porcupine, 'Giraffa' the giraffe, 'Lupa' the wolf, have their own territory within the city, their own Contradal church, their own patron saint, their own museum, clubhouse, uniforms, songs and traditions, so much so that they constitute veritable little cities within the city. Competition between them is intense especially during the period of the Palio. Its intensity varies between the different Contrades, some of which are traditional allies while others are traditional enemies. Hostilities between such Contrades as 'Lupa' and 'Istrice' are so intense that a member of one Contrade would hesitate to venture into the territory of the other.

The most striking thing about the Palio is the extraordinary enthusiasm it arouses among all the citizens of Siena regardless of their age or social and economic status. The local papers talk about practically nothing else all year round, while, during the week of the Palio, multi-coloured special issues are on display everywhere. On the actual day of the event the entire population of the city gathers in the Piazza del Campo. The rest of the city is deserted.

The rejoicings afterwards are quite extraordinary. The whole city resounds with the shouts and songs of the victors who parade through the streets in their medieval costumes until the early hours. That night the whole city is invited to drink in the winning Contrades' clubhouse and everybody is encouraged to visit its museum in which are triumphantly displayed all the trophies of the past.

The institution of the Palio makes of Siena the closest approximation in the modern world to a socially ideal city. All aggression is systematically directed into ritualized channels which are not only harmless but which serve to create an extraordinary feeling of cohesion, and of common Sienese citizenship which overrides all class or age differences. As might be expected, crime and deliquency, drug addiction, alcoholism and all the other manifestations of social disintegration are largely absent in Siena even today when they are reaching epidemic proportions in most of our conurbations.

LEARNING TO LIVE WITH IT

Societies, among whom warfare plays an important role, will, to a large extent, adapt to suffering the casualties which warfare, however ritualized, must give rise to.

One can only understand how this can occur if one understands just how closely integrated are the cultural traits making up a society's behaviour pattern. Among head-hunters, for instance, head-hunting is so 'intricately interwoven with the whole social system' as Lorenz puts it, that its abolition could lead to the disintegration of the whole culture, 'even seriously jeopardising the survival of the people.'[20] Cultural traits therefore cannot be judged by themselves but only as part of a cultural pattern.

Consider the Eskimos: during the hunting season, the basic social unit is the family, interpersonal aggression is fairly high and there is occasional loss of life. If one adds to this the hazards of the Eskimo hunter's life in the inhospitable Arctic wastes that he inhabits, then it is not surprising that his society should have to resort to some strategy, among other things, for evening out the number of females to males. This has been achieved by female infanticide. We may not like it but who are we to judge it? It may provide the least disagreeable compromise for ensuring the stability of an Eskimo society.

Among the Jivaros, polygamy plays an analogous role. Without it there simply would not be enough males to go round.

Among the warlike Nilo-Hamites, such as the Masai and the Samburu, warfare is carried out exclusively by the young men in specific age grades who together are referred to as the 'moran'. They live outside the village in promiscuity with the unmarried girls of the same age. However, they are not allowed to marry and have children. Only the elders, who are no longer of fighting age can do so. This strategy limits the social disruption caused by casualties incurred during raiding expeditions on the neighbouring tribes. When one considers it, it is barbaric to send to war, as we do, men who have wives and children to look after.

REDUCING AGGRESSION

We have seen that there are three strategies that can be exploited for reducing the destructiveness of aggression : fission, ritualization and learning to live with aggression. It is important to realize however that these extraordinarily effective mechanisms are not effective in unnatural environmental conditions, by which is meant those which have diverted too radically from the environment to which a species has been adapted by its evolution. Thus as Harrison Matthews writes,

> When population numbers have overtaken the resources of the environment so that serious overcrowding is brought about, this produces a situation similar to that of animals in captivity where the environment is artificially restricted so that aggression is increased and any chance of escape from the aggressor is denied. In both situations the animals are living in a biologically unsound environment which inevitably distorts the normal patterns of social behaviour.[21]

When the environment changes too radically from that to which a species has been adapted phylogenetically or to which a society has been adapted culturally, then adaptive behaviour, which must include that which reduces that destructiveness of aggression, can no longer be mediated.

In the case of a human society, the same thing happens, if the informational pattern, which is organized in its culture, is interfered with by alien influences such as those exerted by western colonialist powers; centralized governments imitating the industrialized nations or by large commercial enterprises.

The reason for this is obvious. Consider two people sitting down to a game of chess. The principal *sine qua non* for a successful game is that both contestants should know the rules and accept them. So it is with the strategies for reducing the destructiveness of aggression. Since they affect people's lives much more deeply than a game of chess is likely to, people must be taught the rules from their earliest youth and inculcated with the belief that they are fair, just and necessary. This is what happens as children are educated or socialized within a traditional society.

The rules they learn provide them with an integrated pattern of instructions which all give rise to a correspondingly integrated behaviour pattern among which should figure the various strategies for reducing to a minimum the destructiveness of aggressive behaviour.

Experiments with monkeys as well as experience with humans reveal that individuals deprived of a satisfactory upbringing, who have not been subjected to the normal socialization process will find it difficult, if not impossible, to accept a position in a social hierarchy.

As a result, the normal cultural mechanisms ensuring the ritualization of aggression are not operative. All individuals are on their own, neighbours are their enemies and they fight and bicker for every petty advantage they can thereby obtain. They may also display gratuitous violence, probably as a pathological response to the social deprivation from which they must inevitably suffer.

Unfortunately, a society's cultural pattern can only remain intact under specific conditions. For instance, it must not be subjected to environmental conditions that are changing too rapidly or too radically, otherwise the control mechanisms ensuring adaptation simply cannot cope. A society must not expand beyond a certain size, as there is a limit to the extendability of the bonds which hold it together. Also, its culture must not be modified by extraneous influences such as those exerted by missionaries, colonialist educational authorities, etc.

Unfortunately as a society 'progresses' through the various stages that take it from a hunter-gatherer group to a modern industrial state, so are these conditions ever less well satisfied, while it develops new features which tend to increase rather than decrease the destructiveness of aggression, sometimes in a dramatic way.

This is well illustrated in a study by Otterbein of warfare among the Zulus during different phases of their history in the nineteenth century.

ZULU WARFARE

As is generally known the Bantu tribes moved southwards into what is now South Africa between the sixteenth and the nineteenth centuries. Among these tribes were the Nguni, who were shifting cultivators and cattle raisers. The unit of social organization was kept small by rivalry between the sons of a tribal leader which often led to fissions and also by the very nature of the economic system which required a nomadic way of life and which could not support large groups of people living together. Conflicts between tribes were settled by combat, which above all, took the form of a duelling battle. The main weapon used was the javelin, while each warrior protected himself with a shield. The object of war was to obtain prisoners and cattle, the former to be ransomed, the latter to be permanently retained. Pursuit of a fleeing enemy would stop as soon as the pursued dropped their spears. 'This was a sign of surrender and no more blood would be shed.'[22] Since wounds were seldom fatal, the number of casualities was low.

All this began to change by the beginning of the nineteenth century when populations started to expand. There was no longer enough empty land to permit the continual fission of tribes and also to maintain shifting agriculture. As Gluckman writes,

> It became more difficult for tribes to divide and dissident sections to escape to independence; as the Nguni cultural stress on seniority of descent and the relatively great inheritance of the main heir caused strong tensions in the tribes, chiefs began to press their dominion not only on their subordinate tribal sections, but also on their neighbours. The development of this trend was possibly facilitated by the unequal strength of the tribes.[23]

As a result powerful tribes started subduing smaller ones, creating larger social units.

The most successful tribe was the Mtewa whose chief Dingiswayo was a particularly talented general who was able to achieve military success by organizing the age grades of warriors into regiments of soldiers and hence increasing organization and efficiency. Between 1806 and 1809 he established himself as chief of over thirty tribes. One such subject tribe was the Zulu and it was Dingiswayo who was instrumental in arranging for Shaka, the illegitimate son of a Zulu chief, to become their chief.

Shaka had been in touch with the Boers who very much influenced his way of thinking. He totally changed Zulu military tactics, introducing the short stabbing spear which Dingiswayo disapproved of because of the greater number of casualties occurring between armies equipped with it. He discarded the sandals that had hitherto been worn by Zulu warriors, in order to gain greater mobility. He also arranged his soldiers in a 'close order' shield-to-shield formation with two 'horns' designed to encircle the enemy or to feint at his flanks, the main body of troops at the centre and the reserve in the rear, ready to exploit the opportunities of battle.

From 1819 to 1822, Shaka fought a series of wars which led to the creation of an empire comprising no fewer than three hundred different tribes occupying 80,000 square miles of territory, almost the size of the British Isles. During the last four years of his life the empire ceased to expand and military activities were limited to long range campaigns in search of plunder, mainly cattle.

In examining the evolution of Zulu warfare it is essential to look first at what was the goal of warfare at each particular stage of development. It is only when we know the goal that we can understand the military equipment and tactics used and the actual nature of the conflict. During the first stage, conflicts are best regarded as duelling battles fought mainly to obtain bounty. Spears were used only as projectiles. Casualties were slight especially as fleeing enemies were spared. During the second stage, conflicts were 'battles of subjugation' the object being to subdue neighbouring tribes. The army became more efficient, but casualties were still slight.

During the third stage, conflict evolved into battles of conquest whose goal was the creation of an empire. Tactics and weaponry changed, and casualties became very high. During the fourth stage, the goal 'was to keep the army busy rather than conquering new peoples', and casualty rates became low again. What is particularly important is that 'these four types of war correspond to a different level of socio-political development'. In terms of Service's taxonomy, 'duelling battles' occurred, on the tribal level 'battles of subjugation' led to the development of chiefdoms, 'battles of conquest' brought about the emergence of the state, and with the eventual development of empires, long range 'campaigns' became the dominant form of war.[24]

ARTIFICIAL STATES

In our efforts to bring the other countries of the world into the orbit of our industrial society, that we may more easily persuade them to part with their resources and buy our finished products, we have everywhere forced ethnic groups to join together into totally artificial political units on the western model. Co-operation on the part of such groups whose principal relationship is often one of mutual hostility, being difficult to achieve, the tendency has been to put them under the domination of one such group—the most 'advanced', hence the one most easily inculcated with the industrial ethic.

These purely artificial nation-states tend consequently to be empires in disguise: Pakistan being the Punjabi Empire; Nigeria that of the Hausas; Kenya that of the Kikuyu, etc.

In the meantime the age-old rivalry between tribal groups is simply directed into different channels (more modern ones), and it expresses itself clandestinely, more bitterly and more destructively. At the same time whichever ethnic group by whatever means, and however temporarily, is able to control the shaky apparatus of government is consecrated in its legitimacy by international law. Its regime is automatically bolstered by the support of multinational companies, by the governments of industrialized countries and by United Nations agencies, all anxious to protect their investments and expand the sphere of their economic and political influence.

Such an unstable situation can only lead to constant civil wars which, in fact, is what is happening throughout the Third World. But we have done worse than this. We have actively promoted the large-scale movement of populations from one country, even one continent, to another, usually to satisfy the labour requirements of plantations and other industrial enterprises.

Wherever this has happened, whether it be in East Africa, Guyana, Malaya or Singapore, it has led to increasing tension between the original ethnic groups and the new ones. The reason is that they have had no previous experience of each other and have not been able to develop different symbiotic cultural patterns—a *sine qua non* for peaceful coexistence within the same territory. At the same time, the monolithic nation-state of which they are part does not offer them the possibility of territorial separation. In these conditions such countries must constantly be on the verge of civil war, unless the original inhabitants are exterminated or reduced to the status of a subject people.

Our influence has also led everywhere to terrible population pressures, causing people to migrate to less populous lands where tensions are inevitably created with the local inhabitants. In this way the Nepali have been migrating in large numbers to Sikkim, where they now outnumber the local population and have recently upset the regime itself, subordinating the local people to their rule.

Alternatively, white settlers have set themselves up as a dominant cast of administrators and land-owners as happened in Rhodesia (now Zimbabwe), where they gradually eroded local cultural patterns, the morale and the self-esteem of their subject peoples the Mashona and the Matabele.

In other areas, such as India, Ruanda and Burundi, we have upset existing culturally determined arrangements between distinct ethnic groups, that previously permitted them to coexist symbiotically in the same territory. These groups are now competing for the same ecological niche. Warfare has broken out between them and it is but a question of time before one group or the other has been effectively eliminated.

MUTUAL DEPENDENCE OF SOCIETIES

One of the obvious causes of war must be a resource shortage of some sort, making it desirable for one country to acquire a resource in the possession of another. Among hunter gatherers this is very unlikely to occur. As already mentioned, they tend to have all the land they need, use very few resources and those which they use are easily provided by the environment in which they live, especially as they live off the interest and not the capital of available resources.

This is also largely the case with traditional agricultural societies. These will probably indulge in trade with neighbouring peoples, but this is likely to involve their surpluses only, their dependence on foreign goods being minimal.

We have seen how Britain has reacted to a resource shortage. Fish is getting scarce, the Icelanders are making it scarcer, at least to the British, so out goes the Royal Navy. Precisely the same thing is happening in the Mediterranean. The Moroccans are threatening Spanish fish catches and the latter are threatening violence to ensure their supplies (1975).

Water, like fish, is also becoming a valuable resource. In Britain there is already talk of rationing it, and it is impossible to satisfy the US's ever increasing needs for water without importing it from Canada. Are the Canadians willing to see their lakes plundered for the benefit of American industry? There is already considerable opposition and it is likely to grow. If it becomes too powerful what will the Americans do? Allow the economy to collapse or invade Canada?

A few Arab leaders now have it in their power to cut off the West's oil supplies—which provide an ever increasing proportion of our energy requirements. What would the US do if this were really to occur? Senator William Fulbright, Chairman of the Senate Foreign Affairs Committee, actually stated in the Senate during the oil crisis in 1973 that American policy-makers 'may come to the conclusion that military action is required' to protect American oil interests. Needless to say, there was an outcry. Highly cherished illusions were shattered. America, like Britain, likes to think of itself as a benign and peace-loving country whose sole preoccupation is the happiness and welfare of the people of the earth. Now we are told that she should actually go to war, and for what reason? Not to defend some great ideal like 'the American way of life' against those wicked Communists, but simply in the interests of commerce.

Maurice Strong, when he was Executive Director of United Nations Environment Programme, pointed out the new possibility for armed conflict, created by modern technology. Since rain-making is now a possibility, he foresees 'rain stealing' where one country sets out purposefully to steal another's rainfall thereby causing drought. There have already been disputes between countries over claims of pollution damage to waterways and the atmosphere. Such disputes are likely to increase with the further development of large scale technology. Countries could even attempt to melt the Polar ice cap in search of minerals, thereby creating artificial earthquakes and tidal waves.

The US has already been accused of using weather modification in South East Asia to increase and control rain for military purposes.

In the Declaration of Principles adopted at the UN Conference on the Human Environment in Stockholm in June 1972, a declaration, to which all governments represented were a party, pledged governments to ensure that their activities did not cause damage to the environment of others. What action can the UN apply against the infringements of this declaration? The answer is practically none, other than awakening world public opinion against any possible culprits.

The destructiveness of war is also increased by rapid industrialization. Wars were previously waged with primitive weaponry, causing but minor damage whereas now the most lethal modern weaponry is employed, which world powers, eager to extend their sphere of influence and to increase their exports of armaments, are only too pleased to provide.

This often leads to quite ludicrous situations. Thus in the war between the Ibos and the Hausa the former were armed by the Chinese, the French and the Czechs, whereas the Hausa obtained their armament from the Soviet Union, Egypt and the UK.

VULNERABILITY

As industrialization proceeds, as society becomes increasingly dependent on technical devices for its physical sustenance, so must it become correspondingly more vulnerable. Consider the vulnerability of a large modern conurbation: it depends for its fresh water supplies on complex sewage and purification plants: for its food on increasingly distant agricultural enterprises, elaborate processing plants, and a network of roads and railway lines. It requires a vast bureaucracy to provide the welfare services that prevent its inhabitants from succumbing to all sorts of material and psychological deprivations, a massive national health service with hospitals, surgeries, pharmaceutical laboratories and distributors to prevent them from succumbing to disease, an increasingly large police force, a fleet of armoured cars, vast numbers of safes, burglar alarms and other devices to protect them from the depredations of an ever expanding criminal class, and an increasing number of factories, office buildings and department stores, to provide them with employment and the basic necessities of life.

The whole of this precarious edifice could collapse like a house of cards in the face of strikes, sabotage, or simply as a result of a shortage of some key resource such as energy.[25]

With the accumulation of nuclear weapons, by an ever larger number of different governments, the possibility of their accidental use, if nothing else, must increase correspondingly.

Unfortunately if there is a finite possibility that something will happen, then it must be but a matter of time before it actually does. Accidents are tolerable when they are on a small scale, not so however when they occur on a scale that can lead to the annihilation of whole populations.

Apart from the danger of accidents there is also the danger that nuclear devices might be stolen by terrorist groups. With the development of breeder reactors, the amount of available plutonium will increase considerably. The possibility that some of this lethal substance might be acquired by irresponsible groups is a very real one, and once it is available, the production of nuclear devices by such groups becomes relatively feasible.

Armed conflict is also likely to be favoured by the rise to power of dictators who are in a position to flout public opinion and act on personal impulse.

It is important to note that dictators do not exist in traditional societies.[26] Such societies are governed by public opinion, which in turn reflects traditional values. The Council of Elders, which is the closest thing they have to an institutionalized government, does little more than interpret traditional usage. Kings and Queens, even divine ones, are subjected to all sorts of social constraints which eliminate the possibility of arbitrary action outside the most limited sphere of their personal relationships.

VICIOUS CIRCLE

Unfortunately, once a country takes on aggressive stance it becomes very difficult for it to extricate itself. Apart from not being able to take the risk that a rival power could obtain an advantage over it, the country is faced with another equally daunting problem. If the arms race came to an end, there would be very serious unemployment.

Already arms production is used as a means of bolstering employment. Thus, as Mary Kaldor points out ' . . . one of the reasons given for the fact that the military version of the VC10 costs £1.4 million per unit more than the civil version, was the extra cost of £366,000 per aircraft involved in a policy decision to subcontract work in Northern Ireland. In Italy, most defense contracts specify that 30 per cent of the work must be done in the South.'[27]

As long ago as 1974, Westlake noted that:

> Nuclear disarmament would, of course, release thousands of scientists and engineers, and tons of materials. It has been estimated that 50 million people—more than the whole working population of Britain and West Germany—are engaged directly or indirectly for military purposes throughout the world. By 1970 the armed forces absorbed 23 to 24 million people.[28]

Since this chapter was written, the role of armaments in bolstering up the economy and providing business to powerful industrial interests has become even more apparent. This is particularly clear in the case of the Strategic Defense Initiative or Star Wars programme in the US. The former Secretary of Defense James Schlesinger considers that the present plan will cost as much as a trillion dollars. This would make it in the words of William Hartung, 'the leading area of growth for the nation's major military contractors in the 1990s.'[29] The latter he explains badly need a programme of this sort 'to fill a financial vacuum that will be left when contracts for production of the B-1 bomber, MX missile, and the Cruise and Pershing missiles run out over the next five years.' It is seen strategically as an alternative to the much touted nuclear freeze which was unacceptable to the business community since, if implemented, it would have reduced economic activity whereas the SDI has the opposite effect. As Hartung notes 'If the arms industry had been asked to devise the most profitable alternative to arms control, they couldn't have come up with a better proposal than the Star Wars plan.'

In the meantime, as tensions mount, the industrialized countries will become more desperate to maintain exports so as to pay for the increasing social and ecological side-effects of industrial activity.

At the same time they will be increasingly preoccupied with maintaining unemployment at an acceptable level. One can thus expect the armaments business to continue expanding and the nations of the world to be further caught up in the vicious circle of the armaments race, which must continue until such time as global catastrophe, towards which it is inexorably leading, reduces once and for all our capacity to produce instruments of mass death on the present scale.

SUMMARY AND CONCLUSION

Aggression is a normal and necessary feature of the behaviour of social animals including humans. Its function is primarily to establish the status of each individual within a social group. In this way fighting and bickering are avoided and co-operation becomes possible. In this way too the group becomes a society rather than a random collection of individuals. The qualities which will determine someone's position within a social hierarchy will vary according to the ecological niche that society fills (whether it earns its livelihood by fishing for instance or by agriculture).

Aggression between societies serves the purpose of spacing them out, and thereby preventing them from becoming too big, and maintaining their integrity and internal cohesion.

The destructiveness of aggression can be reduced to a minimum in a number of ways: firstly, by fission, secondly by ritualization, thirdly by learning to live with it. The mechanisms involved are culturally determined. They are only operative when a society is living in an environment that closely approximates that in which it has evolved.

As one departs from this situation, as large-scale agricultural societies develop and worse still as industry takes over, so do such controls become increasingly inoperative. New conditions appear which have the opposite effect, increasing the destructiveness of aggression. These conditions are created principally by the increased dependence of societies on each other, the increased capital-intensity of weaponry, the development of authoritarian government and the general disintegration of societies under the stresses of industrialization.

Self-righteous exhortations in favour of peace or pious declarations of the universal brotherhood of man can serve no purpose save to mask the real issues.

The problem can, in fact, only be solved by methodically and systematically de-industrializing and decentralizing society, thereby recreating those conditions in which new cultural patterns can re-emerge, once more to regulate aggression both between individuals and between the societies into which they are organized.

Notes:

1. Frank Barnaby, Megatonomania, *New Scientist*, April 26 1973.
2. Erenius Eibl Eibesfeldt. The Fighting Behaviour of Animals. *Scientific American*. December 1961.
3. James Olds, Pleasure Centres in the Brain, *Scientific American*, 1956.
4. Derek Freeman, Human Aggression in Anthropological Perspective. in J. McCarthy and F. Ebling, eds. *The Natural History of Aggression*, New York, Academic Press, 1966.
5. Adolf Portman, Personal Aggressiveness and War, in E. Durbin and G. Catlin, eds. *War and Aggression*, London, Kegan and Paul, 1938.
6. A. Kardiner, *The Psychological Frontiers of Society*, New York, Columbia University Press, 1974.
7. E. Eibl Eibesfeldt. op.cit.
8. Margaret Mead, *Cooperation and Competition Among Primitive Peoples*, Boston, Beacon Press, 1961.
9. Brian McDermott, personal communication.
10. Edward Goldsmith, A Model of Behaviour, *The Ecologist*, Vol.2, No.12, December 1972.
11. Konrad Lorenz, *On Aggression*, New York, Harcourt Brace and World, 1963.
12. E. Eibl Eibesfeldt, op.cit.
13. Desmond Morris, Discussion in J. McCarthy and F. Ebling, eds. *The Natural History of Aggression*, op.cit.
14. J. R. B. Love, *Stone Age Bushmen of Today*, London, Blackie and Sons Ltd., 1936.

15. Andrew P. Vayda, Maori Warfare *in* Paul Bohannan, ed. *Law and Warfare*, New York, The Natural History Press, 1967.
16. Rafael Karston, Blood, Revenge and War among the Jivaro Indians *in* Paul Bohannan, ed, op.cit.
17. Robert Lowie, *Primitive Society*, Routledge, Kegan Paul, London, 1953.
18. J. C. Ewers, Blackfoot Raiding for Horse and Scalps *in* Paul Bohannan, ed., op.cit.
19. G. P. Murdock, *Culture and Society*, University of Pittsburgh Press, 1965.
20. Konrad Lorenz, op.cit.
21. L. Harrison Matthews, Overt Fighting in Mammals, in J. McCarthy and F. Ebling, eds. *The Natural History of Aggression* Op.cit.
22. Keith Otterbein, The Evolution of Zulu Warfare *in* Paul Bohannan ed., op.cit.
23. Max Gluckman, The Rise of the Zulu Empire, *Scientific American*, 1960.
24. Elman Service, *Primitive Social Organisation*, New York, Random House, 1962.
25. Gordon Rattray Taylor, Technology and Violence, *The Ecologist*, Vol.3, No.12, 1973.
26. Lucy Mair, *Primitive Government*, London, Penguin Books, 1962.
27. Mary Kaldor, *European Defense Industries, National and International Implications*, ISIO. 1st series, No.8, 1974.
28. Melvyn Westlake, Arms Bill. Sad comment on man's priorities. London, *The Times*, March 15, 1974.
29. William Hartung, Star Wars Pork Barrel, *Bulletin of the Atomic Scientists*, January 1986.

CHAPTER SEVEN

De-industrializing Society

De-urbanizing society.

De-industrializing Society

IT SEEMS UNNECESSARY to list the ills our world is suffering from or to demonstrate that they are getting worse, or that the measures undertaken by us to combat them are increasingly ineffective. It is important, however, to determine why all this should be so.

The tendency, of course, is to blame our failure on mere technicalities, errors in the implementation of our policies, *not on the policies themselves*, for these are the only ones consistent with our world-view, hence the only ones the society it gives rise to, is capable of providing.

Let us begin by considering the main features of this world-view. Implicit to it is the notion that the world we live in is *imperfect*. In the Middle Ages, the Cathars and other heretical sects also regarded the world as imperfect. There reaction however, was to cut themselves off from it, and live instead in a spiritual world of their own making. We, on the other hand, have set out *systematically to improve it*. By means of science, technology, and industry, we have persuaded ourselves, it can be transformed into a veritable paradise, in which everyone will have at his or her disposal an extraordinary array of consumer goods and ingenious technological devices, and in which vast specialized institutions will deal so 'scientifically' with all such problems as unemployment, homelessness, ignorance, disease, crime and delinquency, which are supposed to have afflicted us since the beginning of time, that these will be eliminated once and for all.

This transformation is referred to misleadingly as development and the direction it is leading us in is referred to as progress. It is thereby not surprising that any problems which arise are ascribed to underdevelopment, and that, to solve them, it suffices to invest in more scientific research, more technological innovation and more industrial expansion, i.e. in more development, to which—needless to say—our society is, in any case, committed.

In other words, rather than interpret our problems *objectively* (which is what science is supposed to do for us), we interpret them *subjectively* so as to make them appear amenable to the only solutions we can provide without radically altering our world-view and the social behaviour pattern it gives rise to—the only solutions which, among other things, are at present *economically viable* and *politically expedient*.

Thus, for instance, we define poverty as a shortage of material goods, which justifies the production of more and more material goods. It does not occur to us that it might be more realistic to regard it as an *aberrant situation in which more material goods are required than can actually be produced*, for then the solution would be to create those socio-economic conditions in which less goods rather than more were required.

Or again, we interpret the housing problem as a shortage of houses, which justifies the building of more and more houses. It does not occur to us that it might be more realistic to regard it as an aberrant situation largely caused by the disintegration of the family, as a result of which, where there were eight to ten people per house, there are now two or three. This latter interpretation would be inconvenient, since though we know how to build houses, industrial society does not provide the means for restoring the integrity of the family unit.

In the same way, we regard the high crime rate as a sign that the police force is inadequate or that it is not sufficiently well equipped. It does not occur to us that it might be a symptom of social disintegration. This is because, though it has been up to now reasonably easy to engage more policemen, build more prisons, and manufacture more armoured cars and burglar alarms, there is no mechanism available to us for creating a sounder society without compromising the achievement of other goals to which we attribute a higher priority.

If our interpretation of these and all the other problems which face our society today was the correct one, then, on logistical grounds alone, one could state unhestitatingly, *that they could never be solved*, and that the future of man was very grim indeed. Fortunately, it is our interpretation that is wrong. Our problems are of a very different order and the correspondingly different solutions are much easier to apply.   Let us look a little more closely at this process of 'development', or more precisely 'industrialization'—its latest phase. Firstly, it is not autonomous. It does not occur in a vacuum as is implied by modern economics. If the world were a lifeless waste, as is the moon, there could be no industrialization. If it has occurred at all, it is that over the last few thousand million years the primaeval dust has slowly been organized into an increasingly complex organization of matter—the biosphere, or world of living things—or the 'real world' as we might refer to it—which provides the resources entering into this process. Industrialization is something which is happening to the biosphere. *It is the biosphere, in fact—the real world*—that is being industrialized.

In this way, a new organization of matter is building up: the technosphere or world of material goods and technological devices: or the *surrogate world*.

This brings us to the second important feature of industrialization: the surrogate world it gives rise to is in direct competition with the real world, since it can only be built up by making use of resources extracted from the latter, and by consigning to it the waste products this process must inevitably generate.

Let us see why this must be so. The actual building up of the surrogate world occurs in three steps: *Firstly*, resources are extracted from the real world, which can only lead to its contraction and deterioration. Thus, to obtain timber, forests must be felled, causing soil erosion, a fall in the water table, the drying up of streams and increasing the incidence and severity of droughts and floods. To obtain other building materials such as stones, or clay for brick making, still more areas must be deprived of their trees and topsoil.

Secondly, so as to build up the surrogate world of cities, factories, motorways and airports, these materials must be organized differently *elsewhere*. Hence, the land must also be deprived of its trees and its topsoil before being covered with materials such as cement and asphalt which are random to the processes of the real world.

Thirdly, this process, like all others, must give rise to waste products. These become increasingly toxic as industrialization proceeds (as synthetics take over from naturally-occurring materials). Unfortunately the processes of the surrogate world, being far more rudimentary than those of the real world, give rise to correspondingly more wastes, and as they are neither arranged in such a way, nor are they of the right sort to serve as the necessary raw materials for the further development of the surrogate world *let alone for the restoration of the real one*, they simply tend to accumulate as 'randomness' *vis-à-vis*, both of these rival organizations of matter.

To illustrate this point, consider a modern city of a million inhabitants. Wolman has likened it to some vast beast with a very specific metabolism. Every day it must take in some 9,500 tons of fossil fuels, 2,000 tons of food, 625,000 tons of water, 31,500 tons of oxygen plus unknown quantities of various minerals while it must also emit, during the same period, some 28,500 tons of CO_2, 12,000 tons of H_2O (produced in the combustion of fossil fuels), 150 tons of particles, 500,000 tons of sewage, together with vast quantities of refuse, sulphur and nitrogen oxides and various other heterogeneous materials.[1]

If the beast is to keep alive, its metabolism cannot be stopped any more than can that of any other beast. This means that the resources must be extracted from somewhere, the wastes released somewhere else. The latter as we saw in Chapter 5 cannot simply be made to vanish. Pollution-control simply consists in diverting them to where they are likely to do the least harm or to dilute them in the atmosphere or in the seas. (The loss during the recycling process, in the case of most materials, is so great that this does not provide any long term solution). Pollution-control, in fact, is only possible when there are few such beasts around, impossible when there are a large number—for then pollution becomes global rather than local, there is nowhere to divert it to, and nothing left to dilute it in.

It must follow that all three steps involved in the process of building up the surrogate world give rise to a corresponding contraction and deterioration of the real one. Economic growth, in terms of which the former process is measured, is thereby biological and social contraction and deterioration. *They are just different sides of the same coin.*

Unfortunately, we are part of the real world not the surrogate one. In fact, we have been designed phylogenetically (and at one time culturally, too) to fulfil within it specific differentiated functions. It would be very naive to suppose that its systematic destruction would not affect us in some way. To understand exactly how, we must consider the basic features of the real world. Unfortunately, these tend to be disregarded by most of today's scientists, who are more concerned with accumulating trivia than in understanding basic principles.

The most basic principle of the behaviour of the biosphere is that it is goal-directed as can be shown to be the case with all the behavioural systems which comprise it. The goal is stability, which is defined as the ability of a system to maintain its basic structure in the face of change—and hence its continuity—or, in the widest sense of the term to survive. Stability is not a fixed point in space-time but a course or trajectory which a system must adopt in order to remain stable. By doing so, oscillations or discontinuities are reduced to a minimum. It can be shown that primitive societies were geared to precisely this goal. The main preoccupation of their members was to observe their traditional customs, and to hand them down as intact as possible to their children, and to their children's children. It is only a very aberrant society such as ours, that is geared to systematic change in a given direction, and one that can survive for but a very limited period of time.

The second aspect of the biosphere which concerns us here, is that it is self-regulating, as are all the systems which constitute it. Control is achieved by detecting data relevant to the system's behaviour pattern, and interpreting them in terms of a model of its relationship with its environment. This, in the case of a society, corresponds to its 'world-view', in terms of which its policies are mediated and monitored. This notion of self-regulation is so important to the achievement of stability, that the two are normally included in the same concept of 'homeostasis'. If self-regulation is impaired and the system comes to be controlled externally (asystemically) by an agent outside the system and random to it, then there is no longer any mechanism for keeping it on its course towards stability.

Self-regulation was indeed a basic feature of primitive societies which were in fact remarkably well governed by public opinion and without the aid of formal institutions. Ours, on the other hand, are increasingly governed by external or asystemic agencies, dictators or vast bureaucracies, which are not subjected to the control of the social-system as a whole, for the latter has largely disintegrated into a structureless mass which no longer satisfies the requirements of a self-regulating system.

A further consequence of replacing systemic controls by asystemic ones, is that the normal relationship between a system and its sub-systems is interfered with. In the case of a social system this gives rise to *motivational problems*.

A self-regulating system displays 'order'. This means that its parts are differentiated to fulfil specialized functions. They are thereby no longer autonomous. Instead they have become dependent on each other to fulfil that constellation of functions required for their common survival. In other words, they must co-operate with each other and behave in that way that will satisfy the requirements of the system of which they are a part, and hence contribute to the latter's stability or survival. This they do, not because pressure is applied upon them to do so by some external agent, but because they have been designed phylogenetically and ontogenetically to fulfil the requisite differentiated functions. It is by fulfilling them, in fact, that their relationship with the various constituents of their environment is the most stable—that they are thereby best adjusted to their environment—that their needs are in fact best satisfied.

The operation of this principle at the level of the family is quite evident. Parents fulfil their normal functions by behaving in a particular way towards their spouses and children and thereby ensuring the survival of the family, because it *is by so doing that they best satisfy their basic physical and psychological needs*. It will be shown that the same principle applies to behaviour within any self-regulating system such as a community or an ecosystem. I refer to this as the Hierarchical Co-operation Principle which could be stated thus: *in a self-regulating system, behaviour which satisfies the needs of the differentiated parts will also satisfy those of the whole*.

This principle no longer applies in a modern industrial society, in which behaviour either satisfies, however badly, the needs of the part or of the whole but never both at once. Hence, Garrett Hardin's famous allegory: 'The Tragedy of the Commons', which simply could not occur in a self-regulating social system.

The reason is that once a society has disintegrated, it ceases to provide its members with the optimum environment, that in which they can fulfil adaptively the functions they were designed for by their evolution and their upbringing. Once the environment fails to satisfy the needs of its members, they will cease to behave in that way which will lead to the stability of the larger system of which they are a part, or what is the same thing, of the environment it provides them with.

In a disintegrated society it is thus very difficult to obtain the co-operation of its members for any enterprise which is not specifically designed to satisfy individual interests which—because of the society's disintegration—now conflicts with those of the society as a whole. We are now, in fact, faced with a *motivational problem* which could not exist in a tribal society which is a self-regulating system. Their co-operation can only be obtained by offering people a financial reward or, if the enterprise in question appears to be too contrary to their immediate interests, by coercion.

In other words, when a society ceases to be self-regulating, its behaviour is no longer based on the exploitation of existing social forces. To ensure its day-to-day functioning, and hence its survival—in the broader sense of the term—it must exploit forces which are external and random to the system (asystemic), and this causes serious problems.

First of all, to do so presents serious logistical problems, for the work involved increases correspondingly. Thus, as we take over an increasing number of functions from the self-regulating mechanisms of nature, such as the control of pests, the fertilization of the soil, the management of water resources and the government of human societies, *our work load increases correspondingly*.

What is more *there is a limit to the workload we can undertake.* Indeed, the signs are that we are already overstretched and that neither available capital, nor physical resources—to mention but two factors involved, can allow us to take over any more of the functions that are still fulfilled by nature's self-regulating mechanisms. As was pointed out in the SCEP report, were we to decide that the pollination of plants by bumble bees was old-fashioned and inefficient and that it would be advantageous for us to assume this function, even were we to have at our disposal the most ingenious technological devices, we could not do so for more than an insignificant period, *for the logistical problems involved would be insuperable.*[2]

To fully understand the extent of these logistical problems, we must consider just how rudimentary are our asystemic controls, when compared with the systemic ones of nature. As already pointed out, the biosphere, or real world, is an organization. The importance of this notion of organization cannot be over-emphasized. Consider that human beings are made up of a small and very unimpressive array of raw materials. They are 80 per cent water and the market value of the chemicals used in their production is not much in excess of one pound. However, three hundred million years of evolutionary research and development have gone into organizing them in such a way that they give rise to highly complex and sophisticated living creatures such as human beings. Now consider how unimpressive would be the most sophisticated human artefact which could be produced from the same materials.

It must follow that when one has at one's disposal a limited quantity of materials—which must be the case at the best of times, since our world is a closed system from the point of view of materials (though an open one from the point of view of energy)—it is incomparably *more efficient to use them in such a way that they enter into the building up of the real world rather than of the surrogate one—and hence that they give rise to systemic rather than asystemic processes.*

THE PROBLEM MULTIPLIER

We have seen that in a self-regulating system, behaviour satisfies both the needs of the part and of the whole—in other words, it satisfies the needs of all the separate interrelated parts of the system. We have seen that this is no longer the case once a system has disintegrated. Responses now satisfy but a single need. Such responses, must nevertheless affect in one way or another, the other different parts of a system. As Garrett Hardin puts it, 'You can't do only one thing', and since the things we are doing are random at best, damaging at worst, to the other parts of the system, they must cause a corresponding number of maladjustments. *Asystemic processes and controls, thereby, have side effects. Systemic ones do not.* All secondary effects are useful, indeed necessary. The former in fact, are *problem multipliers*, the latter *solution multipliers*.

What is more, the maladjustments they all give rise to will tend to be cumulative, for as further asystemic controls are introduced, so must the surrogate world build up correspondingly, which in turn must cause our environment (made up of the shrinking real world and the growing surrogate world) to divert ever more radically from that to which we have been adapted phylogenetically and ontogenetically, and which must thereby constitute our optimum environment.

The same is true with regard to social problems whose incidence is becoming increasingly intolerable, such as crime, delinquency, alcoholism, drug addiction and vandalism. These are the symptoms of social maladjustment, caused by the disintegration of our social environment and the modification of its physical infrastructure, so that they divert ever further from the optimum—that to which we have been adapted phylogenetically, and are capable of adapting to ontogenetically.

It is also realistic to regard the increased rate of extinction of plant and animal species as the symptoms of ecological maladjustment. The corollary of this: the increased incidence and seriousness of population explosions among micro-organisms, leading to plant and animal epidemics (some of which are resulting in their extinction), and among plants and animals, in particular humans and their parasites also fall within this category, all are due to the increasing diversion of ecosystems from the phylogenetic norm, as a result of the ever increasing impact on them of the surrogate world.

Those inculcated with the world view of industrialism, will object that in the case of humans, such maladjustments are not conceivable by virtue of their supposedly limitless capacity for adaptation. As Boyden points out, however, we only cherish this illusion because the term 'adaptation' has never been adequately defined.[3] Systemic processes are adaptive because they actually solve problems. Asystemic ones, in which category we must include the solutions we apply in our industrial society, can only 'solve' one problem at the cost of creating more, and increasingly worse ones. Quite apart from the fact that they do not work, as in evidenced by the fact that everywhere our problems are increasing, the capital and resources required in their application are ever less available. Boyden refers to them as pseudo-adaptations. They treat the symptoms, not the causes, and by masking them render the disease correspondingly more tolerable, thereby serving to perpetuate it. This is particularly the case with our state welfare system. In reality, if the term 'adaptability' be used correctly, then it is clear that humans are no more adaptive than are other animals. In fact, they are considerably less adaptive than many micro-organisms—that is why waging chemical warfare against them is likely to do more biological damage to us than to the target species.

THE SOLUTION

From the preceding analysis it should be clear that the problems facing the world today can only be solved by restoring the functioning of those natural systems which once satisfied our needs, i.e. by fully exploiting those incomparable resources which are individual people, families, communities and ecosystems, which together make up the biosphere or real world. Before we examine how this is to be done, let us first of all look into one or two of the many implications.

Clearly, this involves as it must do, moving towards a stable or steady-state society. This, however has generally been associated with a stationary economy. However, though a society in which capital investment equalled depreciation, and births equalled deaths would be one of material equilibrium, it would not be a steady-state society. The impact of man's activities on natural systems—or 'ecological demand', as it is referred to by SCEP and in terms of which biological, social and ecological costs can be calculated, is *cumulative* (that is over and above the rate of natural recovery). This means that biological, social and ecological costs would continue to rise.

In a steady-state society, capital investment would equal depreciation, births would equal deaths, *but the actual level of capital investment and births would be very considerably lower than at present*—one that gave rise to an 'ecological demand' which could in fact be met, i.e. one that was equal to the rate of natural recovery. In other words, what we must aim for is not growth, but *negative growth, or economic and demographic contraction*.

It may be argued that this is precisely what we are getting at the moment, since the oil crisis triggered off a world recession. Economic growth has, of course, resumed in the West, but the real standard of living has been falling in the USA since 1973. In Africa, per capita GNP has continued to fall each year. Economic contraction, however, must have a very different effect, in a society still effectually committed to economic growth—methodically nurturing those appetites whose satisfaction only growth can procure, from what it would in a society specifically geared to planned economic contraction.

The former situation must give rise to an ever greater disequilibrium, the latter by adapting to inevitable realities, means systematically reducing it.

One of the main barriers to the acceptance of such a programme is that economic contraction is viewed as synonymous with a reduction in 'wealth' and hence, in the 'standard of living'. If we believe this, it is once again because these terms have been so misleadingly defined. They are used as a measure of the benefits provided by the surrogate world, *but not by the real world*.

In our society the value of things is determined by the operation of the Law of Supply and Demand. Things only acquire value by being drawn into the economic system, and then by becoming scarce, for value in economics is *marginal, not average, value*. 'The more there is of a commodity', writes Samuelson on this subject, ''the less the relative desirability of its last little unit becomes, even though its total usefulness always grows as we get more of the commodity. So, it is obvious why a large amount of water has a low price, or why air is actually a free good despite its vast usefulness. The many units pull down the value of all units.'[4]

What is more, the contention by many economists that such 'externalities' as clean air and uncontaminated water can be internalized, i.e. that they can be taken into account within the framework of modern economic theory would only be true if their real value were, in fact, quantifiable in money terms.

Money, however, is not the currency of nature, nor does nature obey the laws of modern economics. The benefits provided by the real world, such as clean air and uncontaminated water, are not obtainable in proportion to the amount of money spent on them. Money would only buy them if our technological controls were, in fact, effective, which we have seen they are not. Even if they were, money would only be useful once these commodities had been so depleted that they had thereby acquired a sufficient economic value to justify the appropriate expenditure on them.

If our notion of 'wealth' is so inaccurate, so is our notion of 'cost'. Only immediate monetary costs are taken into account by our economists. Damage done to societies and ecosystems is not taken into account. Yet these must at some later date become reflected in monetary costs—which explains in large measure the powerful inflationary trends of today (1975).

It is both ironic and frightening to think of the number of critical decisions taken today on the basis of cost-benefit analyses established by people who neither know what is a real 'cost' nor what is a real 'benefit'. In terms of real costs and benefits—those of the real world rather than of the surrogate one—a policy of deindustrialization, it is easy to show, is the only one that can systematically increase our standard of living.

POPULATION AND CONSUMPTION

It is maintained by some that because of our vast population, it is impractical to dispense with the industrial system. How would we feed so many people without industry, it is asked? This notion stems, partly at least, from the error of regarding food availability from the point of view of an isolated industrial country, rather than from that of the world as a whole. An industrial country has so far been able to provide its inhabitants with more food than has been available to those of the so-called developing countries not because it produced more, but because it persuaded the latter to part with their food in exchange for manufactured goods—a process which is unlikely to be practicable for very much longer.

There is no reason to suppose that industrialization permits increased food production, except perhaps in the very short term, for the inputs that it requires, apart from causing environmental damage which largely accounts for the diminishing returns encountered in their use, are in limited supply, and their price must eventually become prohibitive. Already, no more than 10 to 15 per cent of Indian farmers can afford artificial fertilizers. On the contrary, industrialization, i.e. the building up of the surrogate world, is only possible, as we have seen, at the expense of the real one from which our food is derived. The diversion from the latter of land, timber, water (which is particularly significant in tropical countries where it is often the limiting factor on food production) and, of course, labour (in particular in industrial countries), is seriously reducing food production, as has been shown in particular by Borgstorm.[5]

Seen in these terms, it must follow that the number of people the real world, taken as a whole, can support is inversely proportionate to its level of industrialization. In fact, it is partly because of our massive population, rather than in spite of it, that de-industrialization is our only possible course of action.

THE INDIVIDUAL AS A RESOURCE

During the industrial age, it has become one of our principal preoccupations to use human labour as sparingly as possible, partly because it has become so expensive while other factors of production have been unrealistically cheap, partly too because we have come to value leisure, or non-work very highly, and to provide it has become one of the most generally accepted means of increasing a people's well-being.

Leisure, however, is only prized in those societies which share our notion of 'work' as distinct from other normal day-to-day activities. This notion does not exist in traditional societies, in which economic activities are carried out on a scale which makes them appear relevant to the business of everyday living. This tends to be confirmed by the fact that there is no word for 'work' in the language of tribal peoples.

It is important to note that our massive population provides us with a very considerable amount of non-polluting energy at no cost save in terms of the calories contained in the food which it would, in any case, probably consume. The principle that people must be made to work again with their hands and not just press buttons and pull levers, has considerable implications. Let us not forget that most technological devices are needed specifically for the purposes of saving labour. Tractors, for instance, do not increase food production, they actually decrease it by compacting the soil and by depriving it of the dung which would otherwise have been provided by the horses and bullocks previously used in their place.

Herbicides are only useful because they save the labour involved in weeding. Other pesticides are largely necessary because our large expanses of monoculture provide such an excellent niche for pest populations and, because, by having abandoned rotation, the niche has become a permanent one. If we return to polyculture and crop rotation, we would need far less pesticides. If we also included legumes in our rotation and were energetic enough to spread cow dung on the fields rather than consigning it to the nearest waterway, we could largely dispense with fertilizers. In hill areas, agriculture would benefit still more by labour-intensive methods because terracing can only be maintained by human labour, and without it, especially in tropical areas, the soil rapidly ends up at the bottom of the valley. A rough calculation suggests that it would suffice to increase the agricultural labour force in the UK by four or five times, to enable this country to forgo much of the input of machinery and chemicals which have been introduced over the last thirty years.

The consequent reduction in resources, and hence the capital required, would have a solution multiplier effect. It would quite obviously be anti-inflationary. It would also create a vast number of new jobs. What is more, these would be much more stable ones than those available today in capital-intensive, and hence non-sustainable, industrial enterprises. Our food production would also be put on a stabler basis. Its immense vulnerability to discontinuities of all sorts makes it only a question of time before a serious food shortage, if not starvation, acutally occurs in one or more of the industrial countries of today. The same principle applies to manufacturing. By making it more labour-intensive, similar advantages can be obtained.

THE FAMILY AND THE COMMUNITY

If individuals are organized in such a way that they constitute self-regulating family and community systems, their value as a resource increases very significantly.

We have seen that all available labour must be fully exploited, but how do we pay for it? To give people a monetary wage would be no use unless it could be used to purchase consumer goods and services, many of which as we shall see, will have to be phased out. What then would they be working for? Their food and keep? This smacks too much of slavery. If people cannot be given access to consumer goods and products as a reward for their work, they must be given something else in exchange. But what? There is only one possibility and that is satisfactions of a non-material kind. But what sort of satisfactions? The answer is social ones. The satisfactions which are obtained by fulfilling one's functions as a member of a real family and community. This sounds absurdly idealistic in the materialistic world in which we live. It is, however, both *practical and realistic*. In terms of the hierarchical co-operation principle, behaviour in a stable system must satisfy the needs of both the parts and the whole. The former, as we have seen, fulfil those functions which will enable the latter to survive simply because it is in this way that their own needs are best satisfied. We have noted how this is the case in a stable family system and that it is also true in a community. What, however, are the practical implications?

The family in a stable society is an economic, as well as a biological and social unit. Polanyi and others have shown that in a traditional society, there is, in fact, no behaviour which could strictly speaking be termed 'economic' (designed to maximize the return on capital, or labour).[6] The normal business of producing, manufacturing and selling things was done for social reasons such as to fulfil kinship obligations or to achieve prestige, which are people's dominant motivations in traditional societies. Specifically 'economic' behaviour is only a feature of a disintegrated society in which social considerations are subordinated to purely economic ones.

It is by increasing the scale of economic activities more than anything else that the family and the community have been so terribly disrupted. There is every reason to suppose that by reducing their scale so that they can be fulfilled at a family and at a community level, we could ensure the reintegration of the corresponding social systems.

This would have a key effect. *Economic activity would once more be governed by the self-regulating mechanisms which were once built into the behaviour pattern of self-regulating families and communities.* In other words people would no longer have to be paid and hence given access to vast quantities of consumer goods and services, which, as we shall see, will in any case no longer be available; or otherwise cajoled into fulfilling their economic tasks. These will be fulfilled automatically in the same way that people, even today, tend to fulfil those family functions that are required to assure the welfare of their families.

Even in recent times in the rural areas of our own industrial world, the value of work undertaken on a purely voluntary basis, largely as a means of acquiring social prestige, must not be underrated. In the UK many political and administrative functions are still fulfilled at a local level in this way, by unpaid Justices of the Peace, for example, and other local dignitaries. To transfer these functions to a centralized bureaucracy has served no purpose other than to justify at vast cost our misguided notion of efficiency and social justice.

The implementation of such a policy would have a solution multiplier effect of staggering proportions. Co-operation between the different members of a stable family would reduce very appreciably the following expenditures:

Expenditure on domestic appliances and convenience foods, since all family memebers including children would be available to help with the household chores.

Expenditure on housing, since one of the main causes of the housing shortage is the disintegration of the family unit, the number of people per house having fallen very considerably (in the UK by almost four times).

Expenditure on education, since creches and kindergartens would no longer be required, the family being in a position once more to undertake the earlier phases of their children's education. Educational costs would be significantly reduced with the reduction of emotional instability and delinquency, which are closely associated with family deprivation.

Expenditure on medical services, in particular for old people who occupy an increasingly large proportion of hospital beds—not so much because they require medical treatment, but because the disintegration of the family unit means that there is nowhere for them to go.

The cost of transport, let us not forget the enormous cost to the nation in terms of land wasted in building motorways and airports, in terms of the energy used to power our elaborate transport system; in terms of ecological damage done by the movement of livestock as well as people from areas where certain diseases are endemic, to other areas where the population—human and non-human—have not developed any resistance to the micro-organisms involved and vice versa.

Expenditure on government, quite apart from the prohibitive costs of bad government, there is every reason to suppose that the only effective form of democracy is participatory rather than elective. It is an illusion to suppose that a society in which people's political obligations are limited to voting once every five years is self-governing except in name. Participatory democracy, in which all adult citizens take an active part in running their own affairs, is a different matter. This is only possible in a small community in which there is constant contact with other people, and in which public opinion is formed by the same cultural influences.

THE ECOSYSTEM

The restoration of ecosystems to their original glory is quite clearly impossible. The impact of the world's massive population even with a radically lower level of consumption would be too great. To recreate its principal features is another question. To do so would radically reduce, among other things, expenditure on, and damage by, pesticides of all sorts.

In an untouched tropical forest, there are massive quantities of insects *but there are no pests*, for insect populations are controlled by self-regulating ecological mechanisms. To exploit the same principles of control in a much simplified agricultural ecosystem, it would suffice to reduce the size of fields, increase the amount of cover, plant many different crops instead of a single one, and adopt rotation. In this way variety, temporal as well as spatial, would be introduced, as in a natural ecosystem and with a similar stabilizing effect.

Massive expenditure on unsatisfactory water development schemes could be replaced by replanting forests mainly in the watersheds of great rivers and along their banks, re-establishing marshlands and water-meadows all of which would increase the soil's capacity to retain water, prevent streams from drying up and assure optimum water levels. The best way of storing water is in the soil. Reservoirs are expensive, and they use up valuable agricultural land. In tropical areas they provide an ideal habitat for the vectors of infectious diseases like schistomosiasis and malaria, and much water is lost from them via evaporation. Also, by raising the water table they tend to give rise to waterlogging and salinity, both of which have seriously reduced agricultural production in many parts of the world.

In general, much of the poverty of the third world is the result of the terrible degradation of the natural environment as a result of deforestation and soil-erosion by wind and water. It can only be cured by a massive programme of re-afforestation and soil conservation, with the adoption of ecologically oriented agricultural techniques, together with the reduction of the scale of human economic activities, all of which would, as a result, have extensive solution multiplier effects.

Socio-economic decentralization is a necessary condition for the reconstruction of the family and the community, but it is not sufficient. These basic units of social behaviour must be given once more the power and responsibility for dealing with problems that they have been designed biologically and culturally to deal with.

As we have seen, the family must be made responsible for dealing with what are specifically family problems. *It must not export them*, for instance by consigning children and old people to specialized state institutions. *One cannot over-emphasize the role that has been played in the general deterioration of the world around us by individuals, families, communities and nations, systematically exporting their problems to each other.* Legislative action, in many cases, would be required to prevent this from recurring, and to ensure that social systems at each level of organization were responsible for the solution of their own problems.

One such problem is population growth—one of the most fundamental causes of the world's present plight. Our only method of dealing with it up to now has been to export surplus population to less populated lands. If this had not been possible, the Malthusian thesis would long ago have been vindicated. Ireland, for instance, has three million inhabitants, and already much poverty and unemployment. Imagine what it would be like today, if she had not exported such a large proportion of her population to the US (there are said to be fourteen million Americans of Irish origin; while half the land area of the world, including Siberia, has been occupied by Europeans).

If we have got away with covering our best agricultural land with asphalt, it is because we have been able to persuade non-industrial countries to provide us with their food (which, they badly need themselves) in return for our largely superfluous manufactured goods. Thirty five per cent of India's exports are agricultural produce. We have thereby effectively exported to them our agricultural land, and so far it is they who have starved, and not us.

Pollution is another problem whose main solution, as we have seen, is to export it elsewhere. A good example is Britain, which prides itself on the fact that SO_2 levels have been falling, but this is only true because very high chimneys have been built which allow it to drift across the North Sea to Scandinavia where it is stunting the growth of crops and trees.

So long as social groups can get away with exporting their problems elsewhere, there does not appear to be any reason for them to make any effort to solve them.

It is also true that the real significance of the problems involved is not evident unless they are reduced to a scale that makes them appear relevant to the business of every-day living to which the preoccupations of the vast majority of people are confined.

This is one of the main reasons why important problems must be dealt with at the lowest possible level. A social group will only take the necessary action to control its population, to prevent the destruction of agricultural land, and to prevent other forms of pollution, when it fully realizes that *if it does not do so, nobody else will; its welfare, indeed its very survival depends entirely on its ability to fulfil these functions as effectively as possible.*

But there is a corollary to this principle. If people are to accept responsibility for the solutions to their problems, these must be of their own creation and not of other peoples'. This means that central government can no longer impose any development plans on a local community, which can seriously affect their lives. If the government wishes to build a motorway or a power station, or if a commercial enterprise decides to put up a factory, this must first be accepted by the communities likely to be affected. It is to be noted that something approaching this system is already in force in Switzerland.

A further implication of the same principle, and one which we are likely to find still less acceptable in terms of the ideas of today, is that people are likely to be unwilling to make the effort required to control their population if, at the same time, more people come from the outside. If they are to be responsible for the measures necessary to reduce their own population, they must also be responsible for those which would increase it. This means that a community must be relatively closed—a principle which runs quite contrary to the most cherished ethical ideals of today. Interestingly enough, this also is largely the case in Switzerland from whose decentralized political system we have so much to learn. It is, needless to say, the case in traditional rural societies in which the village a person belongs to, even if he or she no longer lives there, contributes to providing him or her with an identity, as does the family to which he or she belongs. (In India a man's full name consists of the name of his father, his village, his personal name, and that of his sub-caste.)

If a community is to be an effective social system, then its members must be closely associated with each other in a large number of different ways, so that the bonds required to assure its cohesion can be properly established. What is more, for these bonds to be truly effective, it is likely that they should have developed over a long period, preferably since childhood. An educational system, in a traditional human society (as in one of non-human animals) is designed to assure the socialization of its members. Its object is to communicate to children those values which will enable them to fulfil their functions as members of their family and community (a principle which our educationalists have long ago lost sight of).

A community need not be totally closed, a certain number of 'foreigners' could be allowed to settle but again, as in Switzerland, they would not, thereby, partake in the running of the community until such time as the citizens elected them to be of their number.

It is only in this way that a decentralized society can be created, in which the extraordinary resources provided by self-regulating family communities and ecological systems are fully exploited to satisfy real human needs. The question is, how is the critical transition to such a society to be achieved? The answer is, only by the adoption of a carefully integrated programme, and we must assume, however unlikely it may be that it will be adopted by the government of a major industrial nation.

The programme, as we shall see, will have to be divided into distinct parts. These will all be initiated at the same time, though they will proceed at a different pace as they encounter different degrees of inertia. By its very nature, however, the programme would have to be stretched out over a considerable period of time. One cannot transform a society overnight in an orderly way. In addition, the programme would have to be accepted as a whole. One cannot phase out non-sustainable activities without causing all sorts of problems such as inflation and unemployment, unless at the same time one phases in, to replace them, other more sustainable ones. Nor can one phase in the latter without first phasing out the former so as to free labour and resources for this purpose.

For that reason, it is naive to suppose that a government elected for a five-year period can implement anything more than a patchwork of short-term expedients. It is essential that it obtain from the electorate a mandate to implement at least the first part of the programme. To obtain such a mandate, it must first of all make the electorate clearly aware of the extreme gravity of the global situation and hence of its own national one—which so far, governments throughout the world have systematically played down.

THE PROGRAMME

If the programme is to be fully integrated it must be designed to reverse all the essential trends set in motion by the industrial process. This can be shown to consist of six functionally distinct stages (though it is not suggested that they actually occur in that order, since positive feed-back would have caused them to be constantly affecting each other).

Let us consider the nature of these stages. The first stage is the development of the very specific world-view, whose main features we have already described. As Weber was the first to point out, without such a world-view, first entertained by the non-conformists and in particular the Quakers in England, there would probably have been no industrialization.[7]

A new world-view must replace it. A study of the value-systems of traditional stable societies reveals that though they may vary in many details, their basic features are very similar. In fact it can be shown that, for society to remain stable, a number of basic principles must underlie the world-view upon which is based its stable relationship with its social and physical environment. Let us briefly consider the basic principles underlying the aberrant world view of industrialism, in order to see how they may be modified to give rise to an adaptive and hence stable social behaviour pattern.

Humanism. It is essential to the world-view of industrialism that humans should not be regarded as an integral part of nature but rather as above it, and thereby largely exempt from the laws governing the behaviour of other forms of life on this planet. To justify this, one can postulate a number of abstract entities whose possession by humans is supposed to distinguish them from the other less fortunate forms of life. Thus only humans have a soul, they alone display consciousness, their behaviour is supposed to be intelligent, while that of other forms of life is said to be governed by 'blind' instinct. Only human societies are supposed to be capable of cultural behaviour. Such notions are unknown among stable societies for whom nature is holy and cannot be disrupted without incurring the wrath of the gods. It is by de-sanctifying nature that it has become socially feasible to destroy it, and by sanctifying human progress in its stead that the process has been able to proceed at the present disastrous pace. For humanism we must substitute *naturalism*—respect for the natural world of which we are an integral but only a modest part.

Individualism. Individualism is the notion that a person's duties are primarily to him or herself. It is in keeping with our total ignorance of the nature of the natural systems of which we are a part: the family, the community and ecosystem and of how they are related to each other. For an individual to be a member of a community, his or her behaviour must be subjected to the appropriate set of constraints. A community is an organization.

As such it displays order, defined as the influence of the whole over the parts. This influence is achieved by subjecting the parts to constraints which will limit their range of choices by causing them to become differentiated. Individualism is another word for chaos. It is unacceptable in a stable self-regulating society, as it is in any stable self-regulating natural system. For individualism we must substitute *communitarianism*—the need to subject what may appear to be our individual interests to those of the community and the ecosystem.

Materialism is closely related to individualism. In traditional societies people's goals are largely social. The accumulation of material goods plays no part in the strategy of their lives. Material goods only become necessary when they are required for the purpose of satisfying biological and social needs. Karl Marx was wrong when he referred to religion as the opiate of the people. People have always been religious. Religion is an essential part of their sociability which assures the stability of their social environment. It is not *religion*, in fact, but *materialism* that is the opiate of the people.

Scientism. Scientism is the notion that scientific knowledge can serve as the basis for social and ecological control. Let us not forget that there is no precedent for stable societies based on objective scientific information. Until now they have invariably been based on traditional and very subjective information designed to adapt a particular society to its specific environment, rather than *all* societies to *all* environments. It can be shown that only such cultural information satisfies basic cybernetic requirements.

As this becomes generally realized, so must one develop increasing respect for the information organized into the cultural pattern of remaining traditional societies. This is essential to the task of social decentralization. For scientism, in fact, we must substitute *culturalism*.

Technologism. The notion that there is a technological solution to all our problems is a myth closely associated with scientism, since the only solutions which scientific information can give rise to are technological ones. These, however can play no part in the strategy of nature. We must develop instead a quasi-religious respect for the natural systems that make up the biosphere, whose normal functioning provides the only lasting solution to such problems.

Institutionalism. This myth is also closely related to the preceding ones. If benefits are material and technological, then one must create the optimum conditions in which they can be dispensed. Such conditions do not exist in the home, or in a vernacular community. Therefore institutions are set up to provide them. Ignorance of social and ecological cybernetics leads to the essential self-regulating nature of natural systems being ignored, while it is assumed that their control can be more effectively assured by institutions, i.e. external or asystemic controls. For institutionalism we must also substitute a respect for the self-regulating nature of natural systems—a key component of culturalism and ecologism, i.e. of ecosystems and society (like organisms and families) to themselves.

Economism is the notion that things must be done because they are economic, i.e. so as to maximize the return on capital or on other factors of production. This notion is totally consistent with the others. If all benefits are material, technological or institutional then this must be the means of maximizing them and hence of best promoting human welfare. For economism must be substituted *ecologism*, the notion that things must be done to satisfy not a single end but all the basic, and often competing requirements of the community and its natural environment.

Reform of the educational system would also be required to assure the general adoption of the new world-view. It would have to become considerably more decentralized, and the curriculum would also be changed so that the accent might shift from the random accumulation of data to the acquisition of the cultural information favouring the appropriate socialization process.

The second stage in the implementation of our plan is the development of the technology required for achieving its goals. What is required is a shift—from capital intensive industry to developing the 'appropriate' technology for decentralized living.

The third stage is the transformation of society so that instead of satisfying the requirements of the production-consumption process, it would once more be composed of people who are above all, members of families, communities and ecosystems, and whose behaviour is basically that required to satisfy the requirements of these systems and hence of the larger system of which they are a part, the biosphere. The process will come about automatically as society is decentralized and conditions are created in favour of the restoration of the family, the community, and the ecosystem, at which point economic activities will gradually become subordinated to social ones.

The fourth stage is to reverse the system of capital generation, by means of the production—consumption process. Some capital will undoubtedly be required to finance the early stages of the programme designed to prevent social and economic collapse, and to modify the infrastructure of society in such a way as to favour its decentralization. Slowly the need for capital will be reduced as systemic resources replace asystemic ones.

The fifth stage is the reversal of the process which built up the surrogate world by radically reducing the scale of the production process and producing goods that are ever less destructive to the natural environment. The sixth stage is the disruption caused by the fifth which, as we have seen, we interpret in such a way as to justify technological, material and institutional solutions which means further expanding the production process. This means reducing the scale of technological activities to permit the restoration of the self-regulating social systems which make up the real world, on the basis of whose normal functioning these problems could be solved.

The government should set up a separate department to supervise the implementation of each of the six stages of de-industrialization, in line with the requirements of our programme.

TACTICS

It is unlikely that any government will adopt such a programme until such time as the socio-economic system has still further deteriorated. Its first urgent task would be to prevent the collapse of the economy and also that of the society which has become its appendage. This calls for a considerable investment programme aimed at providing the necessary employment but designed to satisfy, at the same time, a number of other associated ends. Thus this employment must be as labour-intensive as possible—it would probably have to be in any case, since there would not be the resources for capital-intensive employment. Available capital must be put at the disposal of the different ministries dealing with the six different stages of the deindustrialization process *so that their efforts may be* synchronized.

The problem of energy will be a pressing one, though it will decrease as the programme gets under way and our need for asystemic energy will correspondingly be reduced. Investment on increased insulation of houses will be an obvious first step. Another will be the decentralization of power generation. This will involve the dismantling of the grid system and its replacement by small total energy systems (designed to use waste heat for local heating purposes). MacKillop has suggested that 2mw generators producing as much as 4mw of heating capacity and catering for some five hundred homes would be most appropriate.[8] Investment in other marginal energy sources such as windmills, water wheels, and solar collectors will also be justified, regardless of whether such devices are economic initially. Although they would account for only a very small proportion of initial energy consumption they will be capable of producing a substantial proportion of the much lower energy consumption which we must aim for.

The general adoption of these new devices will play an important part in changing attitudes to energy consumption, and to the values of industrialism in general. Already they have become the symbols of the new ecological sub-culture which is fast developing, rather as the Charkha is the symbol of Gandhiism.

A further necessary investment is the separation of domestic from industrial waste. At present human sewage, which with labour intensive agriculture would be systematically returned to the land, tends to be polluted with industrial waste. By avoiding this, composting plants can be set up at village and town level, and compost can be made generally available to farmers and gardeners. At the domestic level the composting lavatory would be introduced to replace that most iniquitous device, the flush toilet, thereby reducing domestic water consumption, making compost available for agricultural purposes, and preventing the pollution of waterways.

PHASING OUT OF LABOUR-SAVING TECHNOLOGY

In many cases, the phasing out of labour-saving technology will have to be undertaken as a matter of urgency, simply in order to prevent the collapse of key services, which can no longer be sustained on the present capital-intensive basis. The cost of the educational system in both the UK and the US has been rising much faster than GNP. In the UK the point has now been reached where it has become seriously short of capital. Predictably the reaction is to reduce the number of teachers rather than abandon the use of the elaborate and totally unnecessary technological equipment (language laboratories, audio visual aids, computers, etc.) with which they have been equipped during the last decades. This extraordinary misguided set of priorities must be reversed. The equipment ' is expendable—*schools without teachers just do not work*.

The inputs to modern agriculture are not only increasingly costly, but their supply is particularly precarious. Shortages of fertilizers and pesticides are becoming increasingly common. Phosphates mainly come from Morocco and could be cut off at a moment's notice for political or other reasons. The oil requirements of modern agriculture are notoriously high. It is variously estimated that between five and ten units of fossil fuels are required in the US to produce one unit of food energy. In the interests of avoiding serious discontinuities, agricultural machinery and chemicals must be phased out as rapidly as possible. This will of course have a solution multiplier effect by creating employment, restoring local communities, and reducing pollution.

These processes will occur anyway as the price of capital equipment becomes prohibitive. It is simply a question of accelerating them and synchronizing them with those other measures which would make available the trained labour force together with the appropriate small-scale technology it would require to replace capital-intensive inputs.

Fiscal measures should be introduced to accelerate this trend. This would include two taxes proposed in *A Blueprint for Survival*.[9]

i) A raw materials tax. This would be proportionate to the availablity of the raw material in question, and would be designed to enable our reserves to last over an arbitrary period of time, the longer the better, on the principle that during this time our dependence on such raw materials would be reduced.

ii) An amortization tax. This would be proportionate to the estimated life of the product e.g. it would be 100 per cent for products designed to last no more than a year, and would then be progressively reduced to zero per cent for those designed to last 100+ years. Obviously this would penalize short-lived products, especially disposable ones, thereby reducing resource utilization and pollution. Plastics, for example, which are so remarkable for their durability, would be used only in products where this quality is valued, and not for single-trip purposes. This tax would also encourage craftsmanship and employment-intensive techniques in general.

In addition:

iii) A transport tax would encourage the use of local products which are also likely to be less energy intensive and of a more renewable nature.

iv) A legal amortization rate for equipment that would have to be phased out as a result of the implementation of this programme could be appropriately raised.

The adoption of these measures would give rise to at least three problems:

1) Producers would experience difficulty in competing with foreign companies using more capital intensive methods. The solution lies in *persuading other countries to adopt the same programme*. This may not be so difficult because its adoption in one country would go a long way towards causing others to do likewise. In any case, a discriminatory import duty could be exacted on the produce of countries which had failed to adopt such a policy.

2) Prices are likely to rise because of the increased cost of production, though this is only true *vis-à-vis what they once were, not what they would have been, had the programme not been adopted*.
3) The consequent fall in output would make it no longer possible to pay people the same wage. Such a situation could provide an opportunity for initiating the phasing out of disruptive social services, to make people realize what these extremely inefficient over-centralized state services actually cost them and their families in terms of taxes they have to pay to finance them and the inflationary pressures they give rise to. It should not be difficult to persuade people to forgo their claims on such services in return for a cash payment which could be but a fraction of the per-family cost (especially in view of the increasingly low quality of the services provided). This payment would partly at least compensate people for their reduced purchasing power. Providing people with more money in this way would not be inflationary, since, by the same token, we would reduce government expenditure by a greater amount.

At the same time, a considerable effort would be made to change the *pattern of consumption* so that money could be diverted from the purchase of capital-intensive goods and services to that of more labour-intensive ones. In an industrial society, consumer products are acquired less for the comfort and convenience they might procure than for reasons of social prestige. This being so, to bring about changes in this pattern of consumption, it would suffice to induce corresponding changes in the determinants of social prestige.

The advertising industry has perfected the art of obtaining the connivance of socially prestigious figures in bringing about changes in consumption patterns favouring the commercial success of particular wares. Their services could be obtained for bringing about, in similar manner, changes to the present pattern of consumption which favour the success of our programme. The changes required are in any case those already under way as part of the growing reaction to the industrial way of life. Let us consider some of them.

In the last few years 'self-sufficiency' has become an 'in' word. More and more people grow their own fruit and vegetables. This trend could be radically accelerated. We could follow the example of Italy where five million urban dwellers still indulge in part-time agriculture. People could be encouraged to acquire small-holdings in rural areas. The current tendency toward favouring large agricultural enterprises fiscally and otherwise could and indeed must, be reversed. If this were generalized, it would have an impressive *solution multiplier* effect by leading people to identify themselves with rural communities.

It would also provide them with a new interest in life, a veritable new goal structure, all of which is of key importance to people lost in the anonymous world of large cities and increasingly deprived of any purpose in life. At the same time, more allotments must be created near city centres. They should be regarded as a high priority land use. More and more space could be allocated to this end, as the infrastructure of industrial society is gradually dismantled. All this could be encouraged in many ways. Gardening and agriculture could play an important part in the curricula of schools, and school time could be allocated to work on allotments or farms.

Also, as a reaction to industrially produced food, there is a growing interest in cooking, a creative and satisfying occupation, which should also figure prominently in school curricula. Cooking schools should also be opened for adults. Cooking skills could figure advantageously among the desirable social accomplishments of our post-industrial society, as indeed they did in the pre-industrial one. Another desirable accomplishment is the playing of musical instruments. An orchestra, however amateurish, makes a greater contribution to a festive occasion than does the most elaborate juke box or hi-fi set.

What is, in fact, required is the *ritualization of economic activity*, in the sense in which aggression can be ritualized—that is to say by chanelling it in those directions which cause the minimum damage to the social and physical environment. This means producing goods and services which are not only labour intensive, and make use of naturally occuring materials, but which have a largely aesthetic and ritualistic appeal, instead of those which are purely utilitarian and which are much more destructive, both directly and indirectly, by usurping functions that should be fulfilled by families and communities.

This principle, needless to say, goes quite contrary to the *utilitarian ethic*, which is so strong in such countries as the UK that anything that is aesthetically pleasing tends to be regarded as immoral. As our programme is implemented, conditions will increasingly favour the 'ritualization of economic activity'.

But what happens to those who are already unemployed and those who might lose their employment, in spite of the measures we would take to prevent it? The only solution that satisfies the other requirements of the programme would be to enrol them into a new organization, which could be known as the 'Restoration Corps'

Its role would be primarily to clean up the mess left by a century and a half of industrialization—restore derelict land, replant hedgerows, restore forests, clean up tips where poisonous waste threatens ground water reserves.

The Restoration Corps would fulfil those uneconomic tasks necessary for the success of our programme, so that normal employment would not be adversely affected. It would be organized into local groups, each one responsible for work in its own home area. This is important, since more enthusiasm can be mobilized for cleaning up one's own locality than somebody else's and life in the Corps must be made as attractive as possible since the financial rewards would be minimal.

All unemployed people would automatically have to join the Restoration Corps, unemployment benefits being altogether eliminated. This is very important since unemployment is not merely a question of material but also of social deprivation, leading to loss of self-esteem, and causing demoralization, broken marriages and social deviancy. In this way the welfare system could be further dismantled.

After graduating from the Restoration Corps, a young person would be made to serve in the Defence Corps, a militia equipped with light weapons only, and organized on a local basis, with periods of duty for several weeks each year, as in the Swiss Army. The present massive expenditure on armaments together with the growing role it plays in international trade, is one of the scandals of our time.

Heavy equipment is unnecessary for the defence of one's homeland. The Vietnam War saw the victory of *people over machines*. It showed that a peasant army, if its morale is high enough, can defeat an army equipped with the most sophisticated weaponry.

Both the local Restoration Corps and the local Defence Corps would help build up local patriotism and the spirit of public service which are quite essential for the effective decentralization of society.

THE PHASING OUT OF CONSUMER PRODUCTS

As the programme is implemented, so would people have less money to spend on consumer products. This means that at the same time the *need for consumer products must be correspondingly diminished*, which, as we have seen, could only be done by restoring the functioning of those natural systems which once provided the benefits for which consumer products provide mere compensations.

One of the most important ways of achieving this is by removing such compensations in as painless a manner as possible. Thus, by reducing the number of consumer goods which people require, it would no longer be necessary for two members of the same family to go out to work.

It is also only in this way that small farms could conceivably survive, for although a small farm can provide a satisfying way of life, *it cannot supply the financial surplus necessary to satisfy today's consumption pattern*. The same principle holds good for artisans and small shopkeepers. The maintenance of today's consumption pattern is equally incompatible with the survival of the ecosystems which make up the real world.

The consumer goods we wish to phase out must simply be *removed from the market*. Taxing them is not sufficient, as they would possibly still be regarded as desirable acquisitions, while their growing inaccessability would lead people to feel that without them their standard of living was correspondingly low. By removing them from the market, on the other hand, life styles would change *to accommodate their absence* and the cost of living would thereby be reduced.

Since consumer goods start off as luxuries and gradually become necessities, as life styles change to accommodate them, we would have to start off by phasing out luxuries which have *not yet been transformed into necessities*. In this category one can include colour television sets, private motor yachts, snow-mobiles, large automobiles, video-tape recorders, electric toothbrushes, electric carving knives, etc. From the point of view of the consumer this is unlikely to cause too great a hardship.

From the point of view of the producer, it would undoubtedly do so if *other activities were not phased in to replace their manufacture*. This, however, presents few problems, in view of the massive new investment programme, described above, in more desirable and sustainable enterprises.

To suggest that dish-washing machines and other domestic appliances should be phased out, would obviously meet with instant opposition. They may be needed in a family consisting of but two or three people and in which both husband and wife must go out to work. They would become quite unnecessary, however, once the family had become re-established and eight to ten people once more inhabited the same house, and also once each family required but a single wage earner for its support. The gradual phasing out of luxury consumer products would have a solution multiplier effect.

In all these ways we would slowly achieve the Gandhian ideal for a nation state as an association of 'village republics' loosely organized into larger social groupings, and in which economic activities were carried out on the smallest possible scale, so as to interfere as little as possible with the social and physical environment.

What is particularly important is that popular enthusiasm should be aroused for the social philosophy which underlies this programme. At present there is considerable disenchantment with the benefits of modern industry, while conventional wisdom is losing much of its credibility. It is but a question of time for this disenchantment to yield to total disillusionment, and for conventional wisdom to become correspondingly discredited in the face of the ever more obvious failure of the expedients it prescribes for solving our worsening problems.

At some point, panic will set in and people will grope about frantically for an alternative social philosophy with an alternative set of solutions. The most attractive is likely to be the most radical—the one which provides the best vehicle for expressing the reaction to the values of industrialism. The ecological social philosophy best answers these requirements.

Right and left wing movements provide but alternative recipes for baking the industrial cake and alternative ways of distributing its slices. We would be offering a totally different cake, the only one whose ingredients are likely to be available, also the only one that can satisfy our real needs and those of society and the world of living things on which we depend for our welfare, indeed for our very survival.

Notes:

1. A. Wolman, The Metabolism of Cities, *Scientific American*, Sept. 1965.
2. Carroll Wilson, *et al. Man's Impact on the Global Environment. The Study of Critical Environmental Problems* (SCEP) Cambridge, Mass, M.I.T. Press, 1971.
3. Stephen Boyden, Evolution of Health, *The Ecologist*, Vol.13, No.8, August 1973.
4. Paul Samuelson, *Economics*, New York, McGraw Hill, 1967.
5. Georg Borgstrom, *Too Many*, New York, Collier-Macmillan, 1973.
6. Karl Polanyi, Our Obsolete Market Mentality, *The Ecologist*, Vol.4, No.6, 1974.
7. Max Weber, *The Protestant Ethic and the Spirit of Capitalism*, New York, Charles Scribner, 1930.
8. Andrew McKillop, personal communication.
9. Edward Goldsmith, Robert Allen *et al*, A Blueprint for Survival, *The Ecologist* Vol.2, No.1, January 1972.

Bibliography

Akpapa, B, Problems of initiating industrial labour in a pre-industrial community, *Cahiers d'Etudes Africaines*, Spring 1973.

Anderson, J. M., *Assessment of the effects of pollutants on physiology and behaviour*, Proc. Royal Soc., London B177, 1971.

Barnaby Frank, Megatonomania, *New Scientist*, 26 April, 1973.

Blodgett, John E., Pesticides Regulation of an Evolving Technology *in* Epstein and Grundy (eds) *Consumer Health and Product Hazards/Cosmetics and Drugs, Pesticides, Food Additives*, Vol.2, The MIT Press, *q.v.*

Bohannan, Paul, ed., *Law and Warfare*, Natural History Press, New York, 1967.

Boyden, Stephen; Evolution and Health, *The Ecologist*, Vol.13, No.8, August 1973.

Boyson, Rhodes; Evidence to the Bullock Committee on Literacy, 1971.

Breis, G. C., UNESCO Courier, June 1972.

Bridges, Bryn; *The Ecologist*, June 1972.

Bunyard, Peter; Gearing up to the Plutonium Economy, *The Ecologist* Vol.6, No.10, 1976.

Burckhardt, Jacob; *The Age of Constantine the Great*, Routledge and Kegan Paul, London, 1949.

Cannon, Walter, B., *The Wisdom of the Body*, W. W. Norton, New York, 1939.

Carson, Rachel; *Silent Spring*, Houghton Mifflin, Boston, 1962.

Carter, Vernon Gill and Dale, Tom; *Topsoil and History*, Univ. of Oklahoma Press, 1974.

Chalmers, W. C., Evidence presented at the Windscale Inquiry, day 48, p.32, 1977.

Coleman, J. S., *The Adolescent Society*, Glencoe Free Press, Illinois, 1968.

Columella, *De Rerum Rustica* (On husbandry), London 1745.

Commoner, Barry, *The Closing Circle*, Jonathon Cape, London, 1971.

Crossland, Janice; Drinking Water, *Environment*, Vol.15, No.3, April, 1973.

Dale, I. M., Baxter, M. S., *et al*, Mercury in Seabirds, *Marine Pollution Bulletin* Vol.4, No.5, May 1977.

de Coulanges, Fustel; *La Cité Antique*, Hachette, Paris, 1927.

Dill, Samuel, *Roman Society in the Last Century of the Western Empire*, Meridian Books, New York, 1958.

Doll, Sir Richard; *Nature*, 265, February 17, 1977.

Donaldson, E. M., and Dye, H.M., Corticosteroid concentrations in sockeye salmon exposed to low concentrations of copper. *Journal Fish. Res. Board*, Can 32, 1975.

Dubos, René; *Pollution*, National Association of Biology Teachers, 1973.

Durbin, E. and Catlin, J. (eds), *War and Aggression*, Kegan Paul, London 1938.

Edsall, J. T., Hazards of Nuclear Power and the Choice of Alternatives, *Environmental Conservation*, Vol.1, No.1, 1979.

Eibesfeldt, Erenius Eibl; The Fighting Behaviour of Animals, *Scientific American*, December, 1961.

Environmental Directorate, OECD Report 1975 and 1976.

Epstein, Samuel; *The Politics of Cancer*, Sierra Book Club, San Francisco, 1979.

Epstein, Samuel and Grundy, Richard (eds), *Consumer Health and Product Hazards/Cosmetics and Drugs, Pesticides, Food Additives*, Vol.2, MIT, Cambridge, Mass, 1974.

Ewers, J. C., Blackfoot Raiding for Horse and Scalps, *in* Bohannan (ed) *Law and Warfare, q.v.*

Fielding, M. and Packham, R., Organic Compounds in Drinking Water and Public Health, *The Ecologist Quarterly*, Summer 1978.

Flowers, Sir Brian; *6th Report of the Royal Commission on Environmental Pollution*, HMSO, London.

Fowler, W. W., *The City State of the Greeks and Romans*, Macmillan, London, 1921.

Freeman, Derek; Human Aggression in Anthropological Perspective, *in* McCarthy and Ebling (eds), *The Natural History of Aggression, q.v.*

Fromm, Eric; *The Art of Loving*, Unwin, London, 1957.

Gibbon, Edward; *The Decline and Fall of the Roman Empire*, Alex Murray, London, 1870.

Glotz, G., *The Greek City and its Institutions*, Kegan Paul, London, 1921.

Gluckman, Max; The Rise of the Zulu Empire, *Scientific American*, 1960.

Goldberg, Edward; *The Health of the Oceans*, UNESCO, 1975.

Goldsmith, Edward; A Model of Behaviour, *The Ecologist*, Vol.2, No.12, December 1972.

Goldsmith, Edward; Blind Man's Bluff. *The Ecologist Quarterly*, Spring, 1978.

Goldsmith, Edward and Allen, Robert, *et al.*, *A Blueprint for Survival*, Penguin Books, London 1972.

Grant, Neville; Mercury in Man, *Environment*, May 1971.

Hammond, Mason; *City-state and World State in Greek and Roman Political Theory until Augustus*, Harvard Univ. Press, 1951.

Harrison Matthews, L., Overt Fighting in Mammals *in* McCarthy and Ebling (eds), *The Natural History of Aggression, q.v.*

Hartung, William; Star Wars Pork Barrel, *Bulletin of the Atomic Scientists*, January 1986.

Heitland, W. G., *Agricola: A Study of Agriculture and Rustic Life in the Greco-Roman World from the point of view of Labour*, Cambrige Univ. Press, England, 1921.

Homo, Leon; *Les Institutions Politiques Romaines*, La Renaissance de Livre, Paris, 1927.

Hughes, C. C. and Hunter, J. M., Development and Disease in Africa, *The Ecologist*, Vol.2, Nos.9 and 10, September and October 1972.

Hume Hall, Ross, *Food for Nought*, Doubleday, New York 1974.

Illich, Ivan; *Deschooling Society*, Calder and Boyars, London, 1971.

Johnson, Anita; The Case against Poisoning our Food, *Environment*, April 1979.

Kaldor, Mary; *European Defense Industries, National and International Implications*, ISIO, First Series, No.8, 1974.

Kardiner, A., *The Psychological Frontiers of Society*, Columbia University Press, New York, 1974.

Karston, Rafael; Blood, Revenge and War among the Jivaro Indians *in* Bohannan (ed) *Law and Warfare, q.v.*

Kemp, Beryl; Nuclear Alert, *Conservation News*, January 1977.

Kiev, Ari (ed), *Magic, Faith and Healing*, The Free Press, New York, 1967.
Kleerekoper, H., Effects of sub-lethal concentrations of pollutants on the behaviour of fish. *J. Fish. Res. Board, Canada*. 32, 1976.
Kreith, Frank; Lack of Impact, *Environment*, January/February 1973.

Lecky, Edward Hartpole; *History of European Morals from Augustus to Charlemagne*, Longmans Green and Co., London, 1905.
Lee, Richard and de Vore, Irven (eds), *Man the Hunter*, Aldine, Chicago, 1968.
Lindahl, P. E. and Schwanbom, E., Rotary-flow technique as a means of detecting sub-lethal poisoning in fish populations, *Oikos*, No.22, 1971.
Linden, O., Acute effects of 081 and Oil Dispersant mixtures on Larvae and Baltic Herring, *Ambio*, Vol.7, No.2, Feb. 1976.
Lorenz, Konrad; *On Aggression*, Harcourt, Brace and World, New York, 1963.
Love, R. B., *Stone Age Bushmen of Today*, Blackie and Sons, London, 1936.
Lowie, Robert; *Primitive Society*, Routledge and Kegan Paul, London, 1953.
Lucretius, *De Rerum Natura*, Penguin Books, London.

Maine, Sir Henry; *Ancient Law: Its connection with the early history of Society and its relation to Modern Ideas*, J. Murray, London, 1861.
Mair, Lucy; *Primitive Government*, Penguin Books, London, 1962.
Malleson, A. *Need Your Doctor be so Useless?* Allen and Unwin, London, 1973.
Marx, Jean; Tumour Promoters: Carcinogenesis gets more complicated, *Science*, Vol.201, 11 August, 1978.
McCarthy, J.D. and Ebling, F.J., *The Natural History of Aggession*, Academic Press, New York, 1966.
McGinty, Lawrence; A Ban on Asbestos? *New Scientist*, 14 July, 1977.
McNamara, Robert; *Development Review*, November 1972.
Mead, Margaret; *Co-operation and Competition among Primitive Peoples*, Beacon Press, Boston, 1961.
Mead, Margaret; *Coming of Age in Samoa*, Morrow, New York, 1928.
Mehta, F. A., *Employment: Basic needs and growth strategy in India*, ILO, Geneva, 1976.
Middleton, John, *et al*, From Child to Adult in *Studies in the Anthropology of Education*, The Natural History Press, New York, 1970.
Murdock, G. P., *Culture and Society*, Univ. of Pittsburgh Press, 1965.

Nisbett, John; Balancing the Costs of Cancer, *Technology Review*, January, 1976.
Norman, Colin and Sherwell, Chris; Treading Softly on the Ozone Layer, *Nature*, Vol.263, 23 September, 1976.

Omo Fadaka, Jimoh; Poverty and Industrial Growth in the Third World, *The Ecologist*, Vol.4, No.2, February 1974.
Otterbein, Keith; The Evolution of Zulu Warfare, *in* Bohannan (ed), *Law and Warfare, q.v.*

Pirenne, Henri; *Economic and Social History of Medieval Europe*, Kegan Paul, London 1937.
Polanyi, Karl; *Primitive, Archaic and Modern Economics*, Doubleday, New York, 1968.
Polunin, Nicholas (ed), *The Environmental Future*, Macmillan, London, 1973.
Powles, John; The Medicine of Industrial Man, *The Ecologist*, Vol.2, No.10, October 1972.

Portman, Adolf; Personal Aggressiveness and War *in* Durbin and Catlin (eds) *War and Aggression, q.v.*

Rahm, O. F., *Chaga Childhood*, Oxford University Press, 1967.
Rappaport, Roy, A., *Ecology, Meaning and Religion*, North Atlantic Books, Richmond, California, 1979.
Rappaport, Roy, A., *Pigs for the Ancestors*, Yale Univ. Press, New Haven, 1967.
Rattray Taylor, Gordon; Technology and Violence, *The Ecologist*, Vol.3, No.12, December 1973.
Reichel Dolmatoff, G., Cosmology as Ecological Analysis: A View from the Rainforest, *The Ecologist*, Vol.7, No.1, January/February, 1977.
Renan, Ernest; *Marc Aurele et la fin du Monde Antique*, Paris 1882.
Rosenthal, H. and Alderdice, D. R., Sub-lethal effects of environmental stressors, natural and pollutional, on marine fish eggs and larvae, *J.Fish, Res. Board*, Canada, 33, 1976.

Sahlins, Marshall; *Stone Age Economics*, Aldine Atherton, Chicago, 1972.
Samuelson, Paul; *Economics*, McGraw Hill, New York, 1967.
Schubert, Jack; The Programme to Abolish Harmful Chemicals, *Ambio*, 1972.
Service, Elman; *Primitive Social Organisation*, Random House, New York, 1962.
Shearer, A. R., *The Economics of the New Zealand Maori*, Govt. Printer, Wellington, New Zealand, 1972.
Shepherd, P. and McKinley, D. (eds), *Environ/Mental*, Houghton Mifflin, Boston, 1971.
Smith, Edwin, W. (ed), *African Ideas of God*, Edinburgh House Press, London, 1950.
Statistics of Education, School Leavers, Vol.2, 1971.
Strouts, R. G., Canker of Cypresses caused by *Coryneum cardinale, European Journal of Forest Pathology*, 1973.
Suetonius, *The Lives of the Twelve Caesars*, Random House, New York, 1931.

Tinker, Jon; Treating Smog with Air Freshener, *New Scientist*, 9 September 1976.
Tucker, Anthony; *The Toxic Metals*, Pan/Ballantyne, London, 1972.
Turner, Victor; A Ndembu Doctor in Practice *in* Ari Kiev (ed) *Magic, Faith and Healing, q.v.*

Wakstein, Charles; British Nuclear Fuel's safety Record, *The Ecologist*, Vol.7, No.6, October 1977.
Waldichuck, Michael; *The Assessment of sub-lethal effects of pollution in the Sea. Review of the Problems.* Paper prepared for the Royal Society Meeting 24/25, May 1978.
Walgate, Robert; Experts agree on Mediterranean pollution treaty, *Nature*, Vol.280, 5 July, 1979.
Washburn, Sherwood, L. and Lancaster, C. S., The Evolution of Hunting *in* Lee and Devore, *Man The Hunter, q.v.*
Weber, Max; *The Protestant Ethic and the Spirit of Capitalism* Charles Scribner and Sons, New York, 1930.
Whisson, Michael; Some Aspects of Functional Disorders among the Kenyan Luo *in* Ari Kiev (ed) *Magic, Faith and Healing, q.v.*
Whitten, N. E. Jnr, Ecological Imagery and Culture Adaptability: The Caneles Quichua of Eastern Ecuador, *American Anthropologist*, Vol.80, No.4, December 1978.

WHO, Chronicle 1973.

Wilson, Carroll, *et al*, *A Study of Critical Environmental Problems (SCEP) Man's Impact on the Global Environment*, MIT Press, Cambridge, Mass, 1971.

Wurster, Charles; The Effects of Pesticides on the Environmental Future *in* Polunin (ed) *The Environmental Future, q.v.*

Wolman, Abel; Ecological Dilemma, *Science* No.193, 1976.

Wolman, Abel; The Metabolism of Cities, *Scientific American*, September 1965.

Yellowlees, W. W., *Journal of the Royal College of GPs*, No.29, 1979.

Zuckerman, Lord; Speech given at UN Conference on the Environment Stockholm. The Environment, *This Month*, London 1972.